Teaching Transformation

Teaching Transformation

Transcultural Classroom Dialogues

By
AnaLouise Keating

TEACHING TRANSFORMATION
© AnaLouise Keating, 2007.

First published in 2007 by
PALGRAVE MACMILLAN™
175 Fifth Avenue, New York, N.Y. 10010 and
Houndmills, Basingstoke, Hampshire, England RG21 6XS
Companies and representatives throughout the world.

PALGRAVE MACMILLAN is the global academic imprint of the Palgrave Macmillan division of St. Martin's Press, LLC and of Palgrave Macmillan Ltd. Macmillan® is a registered trademark in the United States, United Kingdom and other countries. Palgrave is a registered trademark in the European Union and other countries.

ISBN-13: 978–1–4039–7647–5
ISBN-10: 1–4039–7647–3

Library of Congress Cataloging-in-Publication Data

Keating, AnaLouise
    Teaching transformation : transcultural classroom dialogues / by AnaLouise Keating.
        p. cm.
    Includes bibliographical references and index.
    ISBN 1–4039–7647–3 (alk. paper)
    1. Multicultural education. 2. Critical pedagogy. 3. Multiculturalism. I. Title.

LC1099.K43 2007
370.1—dc22   2006052768

A catalogue record for this book is available from the British Library.

Design by Newgen Imaging Systems (P) Ltd., Chennai, India.

First edition: June 2007

10 9 8 7 6 5 4 3 2 1

Printed in the United States of America.

for my teachers and students

CONTENTS

Transformational Multiculturalism: Definitions, Alterations, Interventions

[W]e are in fact interdependent. . . . As human beings, we have a sacred connection to one another, and this is why enforced separations wreak havoc on our souls. There is great danger . . . in living lives of segregation. Racial segregation. Segregation in politics. Segregated frameworks. Segregated, compartmentalized selves. Our oppositional politic has been necessary, but it will never sustain us; while it may give us some temporary gains, . . . it can never ultimately feed that deep place within us: that space of the erotic, that space of the soul, that space of the Divine.

M. Jacqui Alexander[1]

I begin *Teaching Transformation* with these words by Jacqui Alexander because they so closely reflect my motivation for writing this book. Like Alexander, I believe that we share a profound connection that ensures our interrelatedness. This deep-seated commonality, though too rarely acknowledged, offers potentially transformative alternatives to the oppositional pedagogies and politics we generally employ. As Alexander notes, oppositional politics lead to "enforced separations," segregations, and other forms of division that at best bring only partial relief. Based on binary (either/or) thinking and dualistic ("us" against "them") models of identity, these oppositional movements inhibit social change by generating nonproductive conflict, suspicion, competition, and debate. For these reasons and others that I explain in the following pages, I adopt a connectionist approach and posit interconnectivity as a theoretical and pedagogical framework for social change. I borrow the term "connectionist" from

Gloria E. Anzaldúa and use it to describe a nonbinary, creative epistemology. As Anzaldúa explains in "now let us shift . . . the path of
conocimiento . . . inner work, public acts":

> When perpetual conflict erodes a sense of connectedness and
> wholeness la nepantlera[2] calls on the "connectionist" faculty to
> show the deep common ground and interwoven kinship among all
> things and people. This faculty, one of less-structured thoughts,
> less-rigid categorizations, and thinner boundaries, allows us to
> picture—via reverie, dreaming, and artistic creativity—similarities
> instead of solid divisions. (568)

Connectionist thinking is visionary, relational, and holistic. When we
view ourselves and each other from a connectionist perspective, we look
beneath surface judgments, rigid labels, and other divisive ways of thinking;
we seek commonalities and move toward collective healing.

References to healing, "sacred connection," "souls," or other such
topics might sound strange in an academic context such as this book.
When we talk about spirits, transformation, interconnectedness, or the
sacred, we risk accusations of essentialism, escapism, or other forms of
apolitical, irrational, naive thinking that (perhaps) inadvertently reinforce the unjust status quo. As Alexander notes, despite recent scholarship linking spirituality with sociopolitical change,[3] "there is a tacit
understanding that no self-respecting postmodernist would want to align
herself (at least in public) with a category such as the spiritual, which
appears so fixed, so unchanging, so redolent of tradition" (*Pedagogies* 15).
However, as I will demonstrate in the following chapters, holistic, spirit-
inflected perspectives—when applied to racism, sexism, homophobia,
and other contemporary issues—can chip away at and in other ways
transform social injustice.[4]

four premises

Building on this theoretical and pedagogical framework of interconnectivity, *Teaching Transformation* is based on four premises.

Premise #1

*Categories and labels, although sometimes necessary, can prevent us from recognizing
our interconnectedness with others.*

Whether we identify as "of color" or "white," as "female" or "male" or "trans," as "lesbian" or "straight" or "bisexual" or "queer," we have *all* been trained to evaluate ourselves and each other according to the existing labels. We have been indoctrinated into a dualistic worldview and an overreliance on empirical-rational thought, which creates a restrictive framework. This framework marks, divides, and segregates based on narrow, binary-oppositional models of difference. As Patricia Hill Collins explains, "In either/or dichotomous thinking, difference is defined in oppositional terms. One part is not simply different from its counterpart; it is inherently opposed to its 'other.' Whites and Blacks, males and females, thought and feeling, are not complementary counterparts—they are fundamentally different entities related only through their definitions as opposites" (*Black Feminist Thought* 70). This oppositional logic reduces our interactional possibilities to two mutually exclusive options: Either we are entirely the same *or* we are entirely different. In this dualistic either/or system, difference becomes rigid and divisive. When we view the world through this binary lens, we assume that the differences between ourselves and the various others we encounter are too different—too *other*, as it were—to have *anything* (of importance) in common. Where is the room for complexity, compromise, growth, and exchange?

I am not suggesting that we should dismiss all identity categories and declare ourselves from this day forward "color-blind," "gender-blind," and so forth. Absolutely not. My point here is that educators, scholars, and others who are concerned with social justice need to become more aware of how these categories function to prevent us from recognizing our interrelatedness. As I will suggest in the following chapters, we need to think about and reflect on the identity categories we employ; we need to learn and teach their histories; we need to use the categories in new ways.

Social identity categories are dangerous when we use them unthinkingly, without self-reflection. When we *automatically* label people by color, gender, sexuality, religion, or any other politically charged characteristics, we naturalize the categories and assume a false homogeneity within each categorized group.[5] We build walls and isolate ourselves from those whom we have labeled "different." In such instances, the boundaries between various groups of people—and, by extension, the theoretical perspectives designed to represent them—become rigid, inflexible, and restrictive. These categories distort our perceptions, creating arbitrary divisions among us and an oppositional "us"-against-"them" mentality that prevents us from recognizing potential commonalities and working together for social change.

Identity categories based on inflexible labels establish and police boundaries—boundaries that shut us in with those we have deemed "like" "us" and boundaries that close us out from those whom we assume to be different. In this oppositional, difference-based framework, identities become ends in themselves rather than useful tools as we move toward larger goals such as transformation, liberation, and other forms of social justice. When we use identity-based categories in these automatic unthinking ways, the labels function as barriers—*not* as bridges, *not* as pathways but as inflexible, insurmountable obstacles. We trap ourselves within limited worldviews and cannot perceive our interconnectedness with others. Andrea Canaan made a similar point over twenty-five years ago in *This Bridge Called My Back: Writings by Radical Women of Color*: "The enemy is brownness and whiteness, maleness and femaleness. The enemy is our urgent need to stereotype and close off people, places, and events into isolated categories. . . . We close off avenues of communication and vision so that individual and communal trust, responsibility, loving, and knowing are impossible" (236).

Premise #2

Out of all the categories we today employ, 'race' is perhaps the most destructive.

'Race'[6] is, for sure, one of the "master's tools" (to borrow Audre Lorde's well-known phrase[7]), one of the most insidious and divisive tools of all. Categorizing people by 'race' has become an accepted, commonsense way of comprehending and explaining ourselves and our world. In the United States, surveys, census forms, birth certificates, applications for jobs, schools, and scholarships ask us to identify ourselves according to 'race,' and often stipulate that we select only one designation. Indeed, 'race' has become so natural that we rarely (if ever!) question the boundaries of its descriptive power. Even supposedly progressive treatments of 'race' can reinforce this naturalizing process. As Ian Hacking notes,

> Well-intentioned television programming for children constantly emphasizes that the characters, even if they are not human, are of different races. From infancy, children watch television cartoons that show, for instance, a happy black family playing with a happy white family. The intended message is that we can all get along well together. The subtext is that we are racially different, but should ignore it. Experimenters discover that small children expect parents of any color to have children of the same color. ("Why Race Still Matters" 112)

People generally assume that physiological differences such as skin color, hair texture, and facial features between various so-called 'races' indicate distinct underlying biological-genetic and/or cultural differences—differences implying permanent, "natural" divisions between separate groups of people. We have been trained to classify and evaluate ourselves and those we meet according to these racialized appearances: We look at a person's body, classify her, insert him into a category, make generalizations, and base our interactions on these racialized assumptions. These assumptions rely on and reinforce monolithic, divisive stereotypes that erase the enormous diversity within each individual and within each so-called 'race.' As Margaret E. Montoya notes, "Stereotypes are hardwired into all of our brains, and their effects are largely unconscious" (248).

However, as I will explain in more detail in chapter three, this deeply embedded belief in discrete, biologically separate 'races' is scientifically and historically inaccurate—transforming arbitrary, superficial physiological distinctions between people into immutable, "natural," "God-given" facts with highly destructive material and psychic results.[8] Popular beliefs to the contrary, there are no genetically distinct 'races' of people. As James Baldwin, David Roediger, Naomi Zack, Theodore Allen, Thandeka, and many others have demonstrated, racial categories are not—*and never have been*—benign. Racial divisions and the concept of purity on which they rely were developed by those in power to create a hierarchy that grants all sorts of special benefits to some groups of people while simultaneously denying, oppressing, and excluding many others.[9] Racialized categories are built on a series of brutal, exclusionary practices. They originated in histories of oppression, manipulation, land theft, body theft, soul theft, physical and psychic murder, and other crimes against specific groups of people. These categories were motivated by economics and politics, by insecurity and greed—not by innate biological or divinely authorized differences. When we automatically refer to 'race' or to specific 'races,' we draw on and thus reinforce this violent history, as well as the 'white'[10] supremacism buttressing the entire system.

Lest I be misunderstood, let me emphasize: I do not advocate adopting a "colorblind" approach in our classrooms, our scholarship, or more generally in our lives. To do so would distort the potentially profound meaning of "colorblind" and simply reinforce the increasingly popular but very false belief that 'race' no longer matters in contemporary U.S. culture.[11] Recent discussions of a (pseudo)-colorblind culture are diversionary tactics. Subtly reinforcing and normalizing economic, educational, and other disparities, contemporary color-blind rhetoric enables privileged people to deny accountability. Because they maintain that we now live in a color-blind society where 'race' no longer matters, they can reject the need for systemic social change and blame the individual or specific group for their inability to succeed. As Amanda Lewis explains, so-called

> color-blind narratives assert that race is no longer relevant and suggest that any lingering racial inequality is a result of individual or group-level deficiencies (or old problems that will fade with time) rather than a result of current and past racial discrimination. They serve as protection for the existing unequal distribution of resources and as a justification for ending what few programs still exist to try and address racial inequality. (161)

Moreover, as I explain in chapter three, this "colorblind" discourse almost always reinforces an invisible 'white' norm.

And yet, constant references to 'race' perpetuate and reinforce the belief in and the perception of separate peoples, monolithic identities, and rigid stereotypes. This racialized discourse prevents us from recognizing our interconnectedness and working together for social change. As Patricia Williams notes in *Seeing a Colorblind Future*, a "significant part of the failure to see each other is fed by a persistently divided rhetoric of race" (65). The key, then, is to discuss racism and 'race' in complex ways that neither rely on nor become trapped by false assumptions of colorblind sameness.

Premise #3

The oppositional politics so effective in the past are no longer as useful in the twenty-first century; we must develop relational, nonbinary forms of opposition, resistance, and transformation.

Oppositional politics have been crucial, enabling people from oppressed groups and other social-justice actors to seize agency, develop subversive forms of resistance, and redefine themselves in empowering ways. As Bonnie Mitchell and Joe Feagin explain,

> elements of oppositional culture within each non-European group[12] operate to preserve dignity and autonomy, to provide an alternative construction of identity (one not based entirely on deprivation), and to give members of the dominant group an insightful critique of their culture. From this perspective, members of oppressed subordinate groups are not powerless pawns that merely react to circumstances beyond their control but rather are reflective, creative agents that construct a separate reality in which to survive. (69)

Critical pedagogy has both participated in and been deeply informed by these oppositional politics.[13] As such, it offers students and educators new forms of agency and useful tools enabling us to examine and deconstruct the dominating culture's oppressive discourse. However, as I suggested earlier in this introduction, binary logic has in too many ways shaped our oppositional politics, making them no longer as effective as we might believe them to be. Indeed, binary forms of opposition keep us locked within the status quo.

I find at least three reasons for this limited effectiveness. First (and quite ironically) oppositional politics have their source in some of the most negative dimensions of western eurocentric[14] thought and are themselves a tool in oppressive social and epistemological structures. As Flora Bridges notes, oppositional discourse proceeds through battle imagery and energies:

> [W]hat becomes normative, "right," and regulatory within the culture is determined by beating down or stamping out various other alternatives. Norms and values are established by way of domination. In this mental framework the possibility for both/and is destroyed. Both/and thinking is basically determined as irrational, primitive, or illogical. What results is a ravaging, hate-filled dogmatic form of establishing cultural values. (71)

This oppositional "mental framework" leads to inflexible, rigid positions; intragroup conflicts; and judgmental, dismissive attitudes—or what Jacqui Alexander describes as "mono-thinking" ("Remembering *This Bridge*" 98). When we structure our teaching or, more generally, our politics and our lives according to this binary framework and the mono-thinking on which it relies, we assume that there is only one right way to think, act, theorize, self-define, or mobilize for social change. This assumption greatly limits our options on many levels, ranging from the types of alliances we can develop to the forms of social change we can envision.[15]

Second, and closely related, these oppositional energies become poisonous when we direct them toward each other, as we too often do. In such instances, we enact what Timothy Powell describes as "corrosive exchanges" ("All Colors Flow" 168) and embark on a "downward spiral of ever more hostile counteraccusations that tend to irrupt when a multiplicity of contentious and contrasting cultural points of view come into contact" ("All Colors Flow" 175). Although Powell focuses specifically on the hostile debates within academic multiculturalism, I have also seen this dynamic happen many times outside the academy, when people or groups oppressed in similar (not identical) ways attempt to develop alliances that fragment from within, and often over fairly minor issues. Even the most well-intentioned social-justice actors can fall into this intragroup combat.[16]

Our oppositional energies trigger this fragmentation. The us–against-them consciousness we use to resist and transform social injustice fosters oppositionality within ourselves, which we then turn against each other. As Alexander explains in *Pedagogies of Crossing: Meditations on Feminism, Sexual Politics, Memory, and the Sacred*, when we construct

> a life based principally in opposition[,] . . . the ego learns to become righteous in its hatred of injustice. In that very process it learns simultaneously how to hate since it is incapable of distinguishing between good hate and bad hate, between righteous hate and irrational hate. . . . We learn how to hate in our hatred of injustice, and it is these psychic residuals that travel, sometimes silently, sometimes vociferously, into social movements that run aground on the invisible premises of scarcity—alterity driven by separation, empowerment driven by external loss, of having to prove perpetual injury as the quid pro quo to secure ephemeral rights. (325–26)

As Alexander implies, we cannot compartmentalize our lives or even our ways of thinking. The oppositional "hatred of injustice" and the us–against-them stance employed in oppositional forms of consciousness leak into other areas of our lives, infecting how we perceive ourselves and everyone we encounter. When we turn this lens against each other—as we too often do—we implode. We shift our focus from the systemic injustice and battle each other, thus impeding social change and reproducing the status quo.

Third, oppositional politics are of limited effectiveness today because they inhibit our ability to envision alternative possibilities. More specifically, the identity-based issues at the heart of most oppositional politics limit our visions of individual and collective change. Generally, social-justice activists organize their politics and actions around identity-based issues. As Leela Fernandes asserts, "identity continues to serve as the ground from which to work for change and to which to retreat for a sense of safety and belonging" (28). Although this oppositional identity-based approach has been very effective in exposing social injustice and affirming oppressed groups, it cannot bring about radical social transformation. Oppositional movements "cannot produce an alternative future that is free from the very identity-based divisions and inequalities that they oppose . . . because they inevitably must rest on a form of identification that explicitly or implicitly is based on an oppositional distinction from another group" (27). These "oppositional distinctions"—like the underlying epistemology which guides them—keep us locked into existing social conditions, standards, structures, and values.[17]

Premise #4

Radical, liberatory change—on both individual and collective levels—is urgently needed and in fact possible, although not necessarily easy to achieve.

Classrooms are, potentially, places of change. What, specifically, these changes entail is context-specific and open to negotiation. However, no matter what forms these changes might take, they are not automatic; to enact transformation, we need theories that deliberately and self-reflectively move into and out of our classrooms. We need multicultural-feminist theories developed and tested in specific teaching situations and then carefully modified based on student reactions, needs, and concerns. Don't get me wrong: I know that conventional (generally abstract) theory[18] can be important. I enjoy reading it, writing it, and discussing it; but I especially appreciate theory when I can make it useful by applying it to my teaching and, more generally, to my life. And so, in this book I do not simply argue that multicultural-feminist theories can or should be transformational. In addition, I draw on my own experiences to offer concrete suggestions for ways to enact what I call transformational multiculturalism in our classrooms. This phrase, "transformational multiculturalism," embodies my womanist perspective and feminist beliefs, my view of multiculturalism, and a key focus in this book.

I am quite aware of the controversy over multiculturalism. Many radical and progressive scholars reject the term because of the ways it has been co-opted and used to support existing social conditions. Rather than examine racism and other power issues, commodified multiculturalisms (or what I describe later in this chapter as *melting-pot multiculturalisms*) ignore systemic issues by offering facile celebrations of diversity that encourage individuals simply to "tolerate" difference. However, if we just celebrate diversity and tolerate difference, we do not challenge racism, sexism, and other forms of oppression; nor do we expose and transform the underlying systems of power that categorize and rank people in very specific material ways. As Frances Aparicio explains, "As it has been defined, implemented, and contained, multiculturalism (as diversity and as tolerance for difference) bypasses these sites of conflict; instead, it has focused on the 'individual' as the problematic site and has tried to 'correct' . . . discriminatory behavior and verbalizations of prejudice" (578).[19] Vijay Prashad makes a similar point, describing multiculturalism as "the liberal doctrine to undercut the radicalism of antiracism. Instead of antiracism, we are now fed with a diet of cultural pluralism and ethnic diversity. The history of oppression and the fact of exploitation are shunted aside in favor of a celebration of difference and the experience of individuals who can narrate their ethnicity for the consumption of others" (63).

Although I agree with these scholars and others who demonstrate contemporary multiculturalisms' limitations, I am not willing to entirely reject the term. Instead, I choose to nurture multiculturalism's transformational potential and use it in the service of social justice. As I will explain in more detail in the following section, I believe that multiculturalism—when historicized and defined broadly to include dis/ability, economic status, ethnicity/'race,' gender, nationality, region, religion, sexuality, worldviews, and other systems of difference—offers unique opportunities to develop nonoppositional politics, inclusionary frameworks, and broader common ground among diverse peoples and worldviews. When thus redefined and rigorously historicized, multiculturalism enables us to learn from each other's histories, scholarship, and pedagogy. In such instances, both students and teachers can be changed through our exploration of multicultural issues and themes. Although multiculturalism has this potential to bring about change, it cannot do so unless scholars and educators define multiculturalism more expansively, incorporate a more sustained analysis of the dominant-cultural ('white'-supremacist, masculinist) framework, and connect our theorizing more closely to our teaching practices. All too often, discussions of multicultural education focus almost entirely on curricular debates—debates that, perhaps unintentionally, often reinforce the status quo—and ignore the importance of developing new teaching tactics. As Robert Dale Parker asserts, "Arguments for an anticanonical multiculturalism have become so canonical a genre unto themselves that the theorizing of *how* to teach multiculturalism has remained underdeveloped and often even unrecognized as a need" (105, his emphasis).[20]

Because "multiculturalism" has become such a widely used term with multiple meanings, in the following section I provide a brief overview of three general trends in the scholarship, focusing especially on how they support or subvert the dominant culture. Although I illustrate this discussion with examples drawn from studies of U.S. American literature, these three approaches to multiculturalism can be found in other academic disciplines as well.

multiculturalism . . . variations on a theme

"Multiculturalism" is a "vexed term" (Powell, *Unruly Democracy* 7), a "floating signifier" (Hesse) with many meanings.[21] All too often, multiculturalism functions as a happy, uncritical pluralism in which differences are verbally and visually valorized yet in many ways dehistoricized and erased. Commonality—defined simplistically as sameness—replaces difference, as racism, sexism, and other social-justice issues are subsumed under facile celebrations of "diversity" that fit well with contemporary U.S. consumer culture and thus reaffirm the status quo.[22] Applied to literary studies and other forms of academic knowledge production, this melting-pot approach builds on overly simplistic assumptions concerning a common core U.S. culture and posits a distinct set of universal "American" values that it finds in every text it reads; those texts that do not conform are either ignored or rejected as "second-rate."[23] Literature is de-historicized and already-existing standards of excellence are applied to a variety of texts.

By thus positing a literary version of the "level playing field" (i.e., the belief that all U.S. Americans are treated equally and have the same chances and opportunities to succeed), melting-pot multiculturalisms deny the roles power plays in constructing literary values and traditions. Thus, for example, melting-pot multiculturalists do not entertain the possibility that the reason so many canonical nineteenth-century writers were Euro-American men from New England could have far more to do with educational and economic opportunities and with the politics of publishing and canon formation than with enduring standards of intrinsic literary merit.[24] In the classroom, melting-pot multiculturalisms lead to what Robert Stam describes as a "grudgingly accretive" approach, where ethnic-specific texts are occasionally incorporated into syllabi and lesson plans but do not challenge underlying knowledge structures or conventional teaching methods. Educators use the same pedagogical techniques and insist that students apply the same evaluative methods and criteria to these "multicultural" texts. Such modest changes do not alter student perceptions. It is, in fact, often the reverse. The emphasis on an underlying "sameness" reinforces the dominant-cultural belief that, in our post–Civil Rights era, we now live in a "color-blind" meritocracy.

At other times, multiculturalism indicates the existence of discrete ethnic/racial/cultural traditions and groups, or what I call *separatist multiculturalisms*.[25] Like melting-pot multiculturalisms, this separatist approach reinforces the dominant-cultural worldview. More specifically, by referring to the existence of a number of distinct 'races,' separatist multiculturalists insist on a rhetoric of narrowly defined authenticity that supports the "common sense" belief that self-contained social identities are permanent, unchanging categories of meaning based on biology, family, history, and tradition. Applied to studies of U.S. literature, separatist multiculturalists reduce multiculturalism to a series of self-enclosed binary configurations, where each ethnic/racial-specific canon is examined primarily, if not exclusively, in opposition to "mainstream" (eurocentric) U.S. literature. When readers focus almost entirely on the differences between ethnic-specific texts and traditions, they inadvertently reinscribe inflexible boundaries that foreclose possible common ground in the discussion of works by different racialized groups. "Difference" (defined in binary, either/or terms) trumps commonality: though scholars often arrive at profound observations, the literary boundaries they construct prevent us from applying insights acquired within one cultural tradition to others.[26]

Even more dangerously, this exclusive emphasis on difference, coupled with the discourse of monolithic authenticity on which it relies, often lead to assumptions concerning who can most accurately read, write, and analyze culturally specific texts. These assumptions, in turn, lead to turf wars and interdisciplinary battles as various groups attempt to stake out their textual territory.[27] In its most vicious manifestations, readers, scholars, and writers become isolationist and gravitate only toward works corresponding to their own self-perceived identities. Jack Salzman, director of the Columbia University Center for American Culture Studies, makes a similar point in an interview:

What first astounded me here was the lack of interest that people in various ethnic communities have in other ethnic communities. If I do a panel, whether fairly low-keyed or high-powered, let us say on the African American community, the audience will be, for the most part, African Americans. There will be very few Hispanics [sic], very few Asians. If I do a panel on Asian Americans, almost no one who is not Asian American will show up. We use terms such as *discourse* and *dialogue* all the time, but, in fact, there seems to be relatively little interest in dialogue between and among various communities. (Qtd. in Kroeber 56, Salzman's emphasis)

In the classroom, these textual boundaries translate into everyday life, when students assume that the differences between themselves and the various others they encounter are too different—too *other*, as it were—to have anything of importance in common.

Despite their many dissimilarities, both melting-pot and separatist multiculturalisms share a number of traits, including the unexamined, often unspoken belief that 'race,' gender, and (hetero)sexuality are natural categories of identity; a discourse of monolithic authenticity that reinforces and circumscribes these naturalized identity categories; confidence in representation's efficacy; and an unquestioned acceptance of already-existing knowledge systems and beliefs, or what I call *status-quo stories*. As I explain in chapter one, I coined the term status-quo stories to describe the ways commonsense explanations about reality normalize existing conditions and deny the possibility of change.

Less frequently, multiculturalism takes a turn toward the "resistant," the "revolutionary," or the "critical," and investigates underlying relations of power.[28] Unlike melting-pot and separatist versions of multiculturalism, critical multiculturalism intervenes in the dominant-cultural framework and thus represents an important development in multicultural theorizing. Critical multiculturalists reject the commonsense belief that cultural identities are preexisting, organic discoveries. While acknowledging that cultural meaning often relies on what seem to be unitary notions of an authentic past and an essential, self-contained set of unique traits, critical multiculturalists challenge all such narrow assumptions about authenticity. They demonstrate that cultural identities are created, not discovered. As Stuart Hall explains,

> Cultural identities come from somewhere, have histories. But, like everything else which is historical, they undergo constant transformation. . . . Far from being grounded in a mere "recovery" of the past, which is waiting to be found, and which, when found, will secure our sense of ourselves into eternity, identities are the names we give to the different ways we are positioned by, and position ourselves within, the narratives of the past. ("Cultural Identities and Diaspora" 225)

By going "back" to already-existing notions of supposedly self-contained, isolated identities and retracing how cultures have been altered—in uneven, power-inflected, often violent ways—through their interactions with each other, critical multiculturalists unravel, expose, and in other ways critique the 'white'-supremacist (masculinist)[29] framework that naturalizes and universalizes existing social conditions and values. Applied to U.S. literary studies, for example, critical multiculturalism exposes the hidden biases (the eurocentrism, the racism, and other forms of domination) that have shaped existing literary standards. Insisting that it's not enough simply to expose and critique this oppressive framework, critical multiculturalists also seek new forms of criticism, new interpretative methodologies, and new definitions that more accurately reflect the complex realities of U.S. culture.[30]

Like those who champion a critical or revolutionary multiculturalism, I believe that multiculturalism offers vital tools enabling us to interrogate and transform existing knowledge systems and social structures. But I describe the multiculturalism I envision and attempt to enact as *transformational* (rather than "critical" or even "revolutionary") to underscore its potential to effect change on multiple interrelated levels, beginning with ourselves and moving outward to encompass readers, students, and the larger world. Transformational multiculturalism is holistic; it requires nonbinary-oppositional epistemological and pedagogical methods that work in the service of social justice. Interconnectivity is key here, and serves as theoretical framework and justification. Building on critical multiculturalism's insights concerning the relational nature of all cultural identities, transformational multiculturalism enables us to redefine and reconfigure dualistic relationships—whether between "self"/"other," "us"/"them," or "oppressor"/"oppressed"—in nonbinary forms.

Transformational multiculturalism's holistic, connectionist approach leads to new forms of individual and collective agency that make change possible, although not inevitable. Because we are all mutually implicated, albeit in differently power-inflected ways, in the 'white'-supremacist masculinist framework that normalizes social injustice, we can (indeed, we must) destabilize and attempt to change it. Transformational multiculturalism emphasizes both personal and communal agency, complicity, and accountability. In various ways and to various degrees, we co-create this world we inhabit.[31]

Self-reflection plays a vital role in transformational multiculturalism. As I use the term, self-reflection requires intense exploration of both the external world and the inner self. Unlike some forms of self-introspection, which become paralyzing as they focus exclusively on the self,[32] transformational multiculturalism's self-reflection is both outwardly and inwardly directed. This self-reflection is a recursive four-part process in which we become partially aware of our unconsciously held beliefs; explore these beliefs' connections with messages from the external world (parents, media, education, religion, and so forth); investigate the implications of these beliefs; and choose whether to keep them, modify them, or attempt to reject them entirely. In the classroom, this process can be transformative. As Chela Sandoval explains, "mind formed through the imperatives of culture is transformable through self-conscious reflection. It is this possibility of transformation and self-formation that makes the hope for sharing a definition and practice of liberty and justice possible between differing categories of the human" ("Theorizing White Consciousness" 100). Through self-reflection students can investigate the ways their values, perspectives, and beliefs have been influenced by the "imperatives of culture." This recognition de-normalizes the dominant-cultural framework and enables students to become aware of both their own complicity with the unjust status quo and the roles they (can) play in challenging and/or reinforcing it. In the following chapters, I discuss some of the ways I encourage this potentially transformative self-reflection in my students.

As my emphasis on mutual complicity, accountability, and interlocking power-inflected identities implies, transformational multiculturalism does not indicate yet another way to celebrate the diversity of voices and viewpoints in U.S. culture. Nor does it imply one more attempt simply to expand existing knowledge systems by including previously unrepresented or erased peoples, texts, and traditions. Instead, transformational multiculturalism begins with the premise that U.S./American culture has always been multicultural, and explores the reciprocal, uneven movements by which peoples, cultures, texts, and other knowledge productions are altered through their interactions with each other.[33] I describe these complicated, nuanced explorations as *transcultural dialogues.*

I borrow the term *transcultural* from Fernando Ortiz who coined it to describe "the extremely complex transmutations of culture" that occurred in Cuba as peoples from many diverse lands, including (but not limited to) Portugal, England, Spain, Africa, China, and North America came or were forcibly brought to the island (98).[34] Unlike the term *acculturation*, which implies a one-way process in which an individual or group simply "acquir[es] another culture" (98), transculturation represents a multidirectional process in which all individuals and groups are altered through their encounters. These alterations are so profound that they lead to the creation of new culture(s). Transculturation is a difficult, often painful, process; it involves loss and brutal change as cultures meet, conquer, mingle, and in other ways interact. As Ortiz explains, the various peoples who arrived (voluntarily and by force) engaged in mutual encounters,

> always exerting an influence and being influenced in turn: Indians from the mainland, Jews, Portuguese, Anglo-Saxons, French, North Americans, even yellow Mongoloids from Macao, Canton, and

other regions of the sometime Celestial Kingdom. And each of them torn from his [sic] native moorings, faced with the problem of disadjustment and readjustment, of deculturation and acculturation—in a word, of transculturation. (98)

Applied to classroom instruction, transcultural dialogues represent dynamic, synergistic negotiations among differently situated peoples, traditions, categories, and/or worldviews. When we engage in transcultural dialogues, we do not ignore differences among distinct groups but instead use these differences to generate complex commonalities. Through this process of translation and movement across a variety of borders—including but not limited to those established by ability/health, class, ethnicity, gender, 'race,' region, religion, and sexuality—we engage in a series of convers(at)ions that destabilize the rigid boundaries between apparently separate individuals, traditions, and cultural groups. These transcultural dialogues can be intensely painful, for they challenge students to (re)examine and perhaps move beyond the safety of their commonly accepted beliefs and worldviews. As Anzaldúa explains,

Transformation has to go through the body, through the physical, the emotional, the spiritual. Once enacted it disrupts & changes everything from family life to relationships with others on every level (social, job, leisure)—which means confronting the fears of what these changes will bring to inner and outer life both. Transformation is messy, disruptive, chaotic. There's an alchemy to transformation and its stages are for the most part painful.[35]

Transformational multiculturalism's transcultural dialogues can shift our perspectives. Such alterations can, potentially, extend beyond the classroom to impact actions in the larger world. But of course transformation on any level is neither automatic nor guaranteed. It involves careful thought, hard work, flexible goals, and the belief that change is possible.[36]

brief overview

Like many people working in higher education, I received little formal training before I was thrown into a classroom of my own. In one graduate course I learned how to design a syllabus and a lesson plan; however, I did not learn how to teach. I wasn't worried, though! I was convinced that my students were a lot like me: They, too, viewed education as liberatory; they too enjoyed reading, writing, and thinking; and they too believed that education was personally and socially empowering. Thus I assumed that my role as an educator was simply to provide students with interesting and provocative material, encourage them to think for themselves, and facilitate class discussions. Unfortunately, this approach was often ineffective: The majority of my undergraduate students, schooled in what Paulo Freire describes as the banking account model of education, relied on rote learning; they assumed that their task as students was simply to take notes, memorize their professors' assertions, and reproduce this information on papers and tests. For almost their entire lives, they had been *told*—by their parents, religious leaders, and conservative media outlets such as Fox News—what to think. In order to challenge students' passive learning and 'white'-supremacist, patriarchal worldviews, I began to educate myself by reading bell hooks, Gloria Anzaldúa, María Lugones, Henry Giroux, and others. These theorists, coupled with my ongoing experiences in the classroom, enabled me to clarify my goals and develop a transformational theory and praxis of teaching, which employs multiple, interconnected perspectives to challenge students' thinking, action, and worldviews.

Whenever possible, I try to integrate my scholarship with my classroom instruction. Both areas benefit from this interchange. (And, to be honest, for most of my teaching career I have shouldered a twelve-hour per semester work load, often with three very different course preparations and four courses ranging from introductory composition and women's studies to graduate-level English and women's studies courses. Had I not found ways to interweave my scholarship with my teaching, I would not have been able to survive!) My classrooms function almost like a laboratory, where the theory I read and write becomes embodied as I try to translate dense theoretical perspectives into relevant terms and practical contexts. My students are my partners, my co-investigators, in these explorations. Together, we examine a variety of theoretical perspectives and cultural worldviews. We all learn from this endeavor. As they explore how cultural and literary texts are shaped by personal and historical issues, students become holistic-critical[37] thinkers and readers. And, as I attempt to explain and enact the various theories in my teaching, I gain new insights into their strengths and weaknesses. I test new approaches, and I modify these approaches based on my students' reactions and concerns. This combination of theory and practice has shaped my understanding of transformational multiculturalism in crucial ways.

In the following chapters, I discuss some of the pedagogical strategies I have developed as I try to enact transformational multiculturalism.

Each chapter grew from my attempts to negotiate between the theory I read, the students I teach, and the world in which we live. This negotiation shapes each chapter's format, enabling me to interweave personal experience with theory and practice.

Drawing on indigenous teachings and my classroom experiences, chapter one, " 'We are related to all that lives': Creating 'New' Stories for Social Change," explores the theoretical and pedagogical implications of positing interconnectivity as a framework for identity formation and social transformation. Because language, belief, perception, and action are intimately interrelated, the stories we tell ourselves and those we learn from our cultures (our families, schooling, religion, friends, the media, etc.) deeply matter: They influence our beliefs, which in turn affect our perceptions, and these perceptions affect how we act. Our actions, in turn, shape the stories we tell, the stories others tell about us, the ways they perceive us, the ways we perceive ourselves. Yet we rarely reflect on this creative power. All too often, we assume that our perceptions and beliefs accurately reflect the entire truth about reality and ourselves; such assumptions narrow, limit, restrict our worldviews and inhibit our actions. I describe these limiting assumptions as status-quo stories: apparently commonsense explanations about reality that normalize existing conditions so entirely that they deny the possibility of change. Indeed, status-quo stories encourage us to remain the same; they limit our imaginations and prevent us from envisioning alternate possibilities—different ways of living and arranging our lives. Chapter one explores two status-quo stories: 'race' and rugged individualism. After examining the crucial role self-enclosed individualism plays in sustaining racism and other forms of social injustice, I suggest that interconnectivity enables educators to develop transformative pedagogical models.

Chapter two, "Forging Commonalities: Relational Patterns of Reading and Teaching," offers an alternative to difference-based models of multiculturalism and critical pedagogy. I describe this alternative as *relational teaching and reading*. These relational patterns—structured by syllabi, text selection, class discussions, and assignments—begin with commonalities[38] and create pathways into relational investigations of difference, difference defined not as deviation *from* an unmarked norm but as interrelated *with* this norm. I argue that this nonbinary approach to difference, coupled with (sometimes painful) self-reflection, makes transformation possible, on both individual and collective levels. In my classrooms, relational teaching enables me to combine cultural critique with self-reflection. By so doing, I can invite students to examine both their own often unconscious presuppositions and the various sources for and effects of these previously unexamined beliefs. After outlining a theory of relational reading and teaching, I discuss some of the specific ways I have implemented this process in my U.S. literature and composition classes. In this chapter, I argue for the importance of designing situation-specific teaching tactics that acknowledge the ways people of all colors and cultures are variously implicated in historical and contemporary events.

Chapter three, "Giving Voice to 'Whiteness'? (De)Constructing 'Race,' " grew out of my experiences inside and outside the classroom: my attempts to incorporate analyses of 'whiteness'[39] into my teaching and my frustration with recent scholarship on "whiteness studies." The chapter title reflects several trends in contemporary theory. Because these trends involve exposing the hidden assumptions we make concerning racialized identities, they have far-reaching yet rarely acknowledged theoretical and pedagogical implications. The first phrase, "Giving Voice to 'Whiteness,' " refers to academic demands for an analysis of 'white' as a racialized category.[40] The second phrase, "(De)Constructing 'Race,' " reflects my assessment of the dangers in scholars' recent investigations of 'whiteness' and other racialized identities. I use the word "(de)constructing"—with the prefix in parenthesis—to communicate my belief that many progressive, well-intentioned theorists who attempt to *de*construct or in other ways investigate 'whiteness' and/or 'race' often inadvertently *re*construct them by reinforcing the belief in permanent, separate racial categories. Although these scholars generally emphasize the socially constructed nature of racial classifications, they discuss racialized identities in ways that undercut and contradict their insistence that 'race' is a historically changing socio-psychological concept, not an immutable biological fact. Not surprisingly, their work is far less transformative than one might assume. In this chapter, then, I explore the limitations in recent work on 'whiteness' studies and offer a sustained critique of scholars' uses of racialized language.

Building on the previous chapter's investigations, chapter four, "Reading 'Whiteness,' Unreading 'Race,' " describes and attempts to address some of the difficulties that can occur when educators incorporate analyses of 'whiteness' into classroom instruction. When students who self-identify or are labeled as 'white' are introduced to recent investigations of 'whiteness,' they are forced to recognize the insidious roles 'whiteness' plays in U.S. culture. This recognition can trigger a variety of unwelcome reactions—ranging from guilt, withdrawal, and despair to anger and the construction of an extremely celebratory racialized 'whiteness.' In short, they often enact a 'white' backlash, which encompasses a sense of heightened alienation from people identified as non-'white' and the belief that 'white' people are victims of affirmative action and thus the newest oppressed group. Although some self-identified students of colors[41] might find it satisfying to see the 'white' gaze that has marked them as "other" turned back on itself, I question the long-term effectiveness of this reversal because it reinforces the status quo—the belief in separate 'races,' which is itself a key part of the 'white' framework. As I explain in this chapter, I have learned how to manage and at times avoid these reactions by employing what I call "(de)racialized reading strategies," a contradictory yet highly effective classroom practice. (De)racialized reading simultaneously racializes and de-racializes texts. I use this strategy to encourage students to adopt new reading practices that explore how all characters and texts are racialized—marked by name, language, location, and color. Yet the tactics themselves enable me to present these racialized readings in unexpected, temporarily nonracialized ways that destabilize conventional understandings of 'race' by underscoring the potentially fluid, relational nature of all racialized identities.

Focusing primarily on sexuality, chapter five, "Teaching the Other?" challenges the assumption that educators must "be" what they teach. This belief, though generally motivated by the desire not to appropriate or misrepresent the experiences of others, revolves around issues concerning personal identity, experience-based knowledge, and authority, and leads to a number of interrelated questions: Can instructors teach about their sexual/ethnic/gender/class/national "Others," or do all such attempts inevitably appropriate, stereotype, or in other ways silence them? To what extent is authority in the classroom dependent on personal identity and experience? How can we effectively teach "what we're not"? By foregrounding sexuality, this chapter explores a second assumption as well: the belief that ethnicity/'race' constitute the only components of multicultural theory and praxis. Drawing on my own teaching experiences, I discuss six strategies that I have found effective in teaching lesbian/gay/bi/heterosexual/trans material written from a variety of cultural perspectives. These strategies—which include framing each course, fostering intellectual agency, tactical (non)naming, focusing on multiple interlinked issues, questioning conventional definitions, and historicizing the categories—apply to teaching other systems of difference as well. However, because sexuality is often viewed as a highly private, taboo topic that should not be discussed in public (and certainly not in academic settings!), it can be the most difficult classroom terrain in which to maneuver—whether we identify as heterosexual, lesbian, bisexual, or gay.

★   ★   ★

The theories and strategies I describe in this book are based on and developed through my classroom experiences teaching undergraduate and graduate courses in women's studies, composition, and literature at three fairly different institutions of higher learning: an open-admissions medium-sized comprehensive university in eastern New Mexico attended by first-generation college students who identify as "Hispanic" and "white;" a private Catholic liberal arts college in western Michigan attended by first- and second-generation college students classified by others as 'white' but by themselves as "just human;" and a public university in North Texas attended by a racially diverse group of primarily first-generation women college students.

Teaching Transformation is rooted in my own contexts: the three institutions, the specific courses I teach, my own background, and the students I've encountered. I do not assume that my experiences and these tactics automatically apply to your own situations; however, it is my hope that you can adopt, revise, and employ them in useful ways.

CHAPTER ONE

"We are related to all that lives"[1]: Creating "New" Stories for Social Change

Are we just little-brained creatures who wind up with these limiting stories of reality because we can't look and are afraid to listen? Really, what changes the world is the power of a compelling story. But we seem to carefully limit the stories that reach us to those that won't push us to change.

Elana Dykewomon[2]

This ancient idea of relationship must be allowed to arise in our collective consciousness once again. In this perilous world of the twenty-first century, it may well be a matter of our collective survival.

Gregory Cajete[3]

Earlier in my career when I worked in a northern state, I was assigned to a spacious office in a renovated mansion. Built in the early twentieth century, the mansion had a substantial heating system with old-fashioned, steam-generating radiators in each room. My computer work station was in front of a large window and directly above the radiator. I was grateful for the opportunity to appreciate the gorgeous view while working, but I was also concerned about the radiator's effect on the computer (which was fairly new). When the technician visited my office to explain the network system, I expressed my concern about the computer's location: I had been taught that computers should not be exposed to direct heat. And let's face it, I pointed out, we have long, cold winters requiring lots of heat. Shouldn't we move the computer? The young man thought briefly about my question and then dismissed my concern, confidently asserting: "Well, it's *always* been like this. The computer has always been there." He spoke the words with such finality that I knew further discussion would be pointless. Years later, I still marvel at his certainty, his deep faith that the way things are and "always" have been is the way they should be.

This interchange with the computer technician stayed with me. As I sat through several winters, the radiator spewing hot air directly into the computer's hard drive, I pondered the implications of his dismissal. I reflected on similar comments I'd heard over the years:

> "That's just the way things are!"
> "People are just like that. Stop worrying. Stop trying to change them!"
> "Hey! I didn't make the rules! I'm just living my life."
> "Oh well, that's life. Get used to it."
> "Just live and let live."
> "Don't rock the boat."
> "Well, if you don't like it—just leave!"

What certainty. What affirmation of the existing reality. Deeply embedded within these confident assertions are what I call *status-quo stories*: worldviews and beliefs that normalize and naturalize the existing social system, values, and norms so entirely that they deny the possibility of change.

status-quo stories

As the term suggests, status-quo stories reaffirm and in other ways reinforce the existing social system. To borrow Elana Dykewomon's words from my first epigraph, status-quo stories do not "push us to change." In fact, status-quo stories seduce us into resisting change. Status-quo stories limit our imaginations and prevent us from envisioning alternate possibilities— different ways of living and arranging our lives. Status-quo stories train us to believe that the way things are is the way they always have been and the way they must be. This belief becomes self-fulfilling: We do not try to make change because we believe that change is impossible to make. Status-quo stories are divisive, teaching us to break the world into parts and label each piece. We read these labels as natural descriptions about reality.

Status-quo stories contain "core beliefs"[4] about reality—beliefs that shape our world, though we rarely (if ever) acknowledge their creative role. Generally, we don't recognize these beliefs *as* beliefs; we're convinced that these stories are, rather, accurate factual statements about the world. Look, for instance, at how 'race' functions in contemporary U.S. culture. As I explained in the introduction, we have become so accustomed to automatically identifying each other based on skin color, physiological features, and other racialized markers that we generally assume racial categories are permanent, unchanging, and monolithic. Our constant, unthinking, daily references to 'race' make 'race' more permanent and in other ways reinforce its reality.[5]

And yet, the status-quo story about 'race' is far less accurate than most U.S. citizens realize. To begin with, the belief that each person belongs only to one 'race' ignores the many people living in this country who could be classified as "biracial" or "multiracial." Indeed, the implicit belief in discrete, entirely separate 'races' implies a false sense of racial purity, for we could *all* be described as multiracial. As Michael Thornton points out, "there are no such things as pure races" (322). Spaniards, for example, are a mixture "of Black Africans, Gypsies (from India), and Semites (Jews, Arabs, and Phoenicians), as well as Romans, Celts, Germans, Greeks, Berbers, Basques, and probably more" (Fernández 143). Furthermore, the suggestion that we can automatically identify ourselves and others according to 'race' assumes that we are fully cognizant of our ancestry and can posit a straightforward correlation between physical appearance and 'race.' Such assumptions ignore the crucial ways that colonialism, slavery, and miscegenation have impacted and shaped us all. As one of the characters in Pauline Hopkins's *Contending Forces* states,

> "It is an incontrovertible truth that there is no such thing as an unmixed black on the American continent. Just bear in mind that we cannot tell by a person's complexion whether he be dark or light in blood, for by the working of the natural laws of the white father and black mother produce the mulatto offspring; the black father and white mother the mulatto offspring also, while the *black father* and *quadroon* mother produce the black child, which to the eye alone is a child of unmixed black blood. I will venture to say that out of a hundred apparently pure black men not one will be able to trace an unmixed flow of African blood since landing upon these shores!" (151, her emphasis)

Similar comments can be made about people labeled as "Latina," "Native American," or any other so-called 'race.' Appearances can be very deceptive, and not one of us is "unmixed."[6]

The myopic perspective on unmixed, discrete, biologically separate 'races' relies on nineteenth-century pseudoscientific theories developed in support of slavery, colonization, and other aspects of an unjust socioeconomic system. As Kwame Anthony Appiah notes, although contemporary scientists have not arrived at a consensus on what 'race' means, "[w]hat most people in most cultures ordinarily believe about the significance of 'racial' difference is quite remote . . . from what the biologists *are* agreed on." Whereas biologists can interpret scientific data in various ways, they cannot demonstrate the existence of genetically distinct "races," for "human genetic variability between the populations of Africa or Europe or Asia is not much greater than that within those populations" ("Uncompleted Argument" 21, his emphasis).

I will return to this story of 'race' in a later chapter. Here, I want to focus on another deeply embedded, extremely destructive status-quo story: the national celebration of rugged individualism and its pull-your-self-up-by-the-bootstraps theory of success. These status-quo stories of rugged individualism and 'race' are closely related. As john a. powell demonstrates, the development of a modern selfhood—where the self is defined as "separate and isolated"—occurred simultaneously with "the emergence of whiteness and racialized hierarchy" ("Whiteness" 36–37). These concepts, greatly influenced by Locke, Hobbes, Hume, and other seventeenth-century thinkers, still influence us today. My experiences in the classroom have convinced me that this Enlightenment-based story of the "self-made-man" is one of the most damaging stories in U.S. culture.

This highly celebrated individualism is one of the dominant stories of our time and a foundational element of western modernity. As David Theo Goldberg notes, "[t]he philosophical commitment in the tradition of Western modernity is to radical and atomistic individualism—to rational, . . . self-interested, self-maximizing, and self-providing individuals" ("Introduction" 25). In the United States, this commitment to possessive individualism has taken an extremely virulent form. According to this status-quo story, each individual is free to pursue his or her own desires—regardless of how this pursuit might impact others. The belief in a fully independent self has, to a great degree, shaped the ways we see ourselves, other people, and the world. We are trained to view ourselves as entirely self-sufficient and self-directed human beings; each of us is (or should be) fully in charge of and responsible for our own lives. As Harlon Dalton observes, "All of us, to some degree, suffer from this peculiarly American delusion that we are individuals first and foremost, captains of our own ships, solely responsible for our own fates" (105).

I am not criticizing all forms of individualism. I value and understand the importance of personal agency, integrity, relational autonomy, and self-respect.[7] Nor am I setting up a simple binary opposition between the individual and the collective. Rather, I focus specifically on an extreme type of individualism that defines the self very narrowly, in *non*-relational, egocentric, possessive terms. According to this view, each individual has a unitary, unchanging core self that must be nurtured, protected, and in other ways honored at all costs. I describe this individualism as *self-contained* or *self-enclosed* to emphasize its rigid borders, extreme detachment, and absolute isolation: each individual is entirely separate from the external world, with permanent, inflexible boundaries dividing the self from all other human and nonhuman life.

This division sets up an adversarial framework that valorizes and naturalizes competition, self-aggrandizement, and fear. As powell explains, the concept of individualism that developed from Enlightenment thought was deeply influenced by Hobbes's theory of human nature: "This individual was not only separate from other individuals; this individual was threatened by other individuals. So the role of society was to protect that individual from the threat, from the terror, of other individuals; to allow that individual to keep whatever she or he had gained in fair exchange" ("Does Living a Spiritually Engaged Life" 37). It's "Every Man for Himself," to borrow an old sexist adage. And the sexism is quite appropriate here. As a number of feminist and multicultural scholars have observed, our U.S. American versions of solipsistic individualism were developed by and for a very select group of human beings—property-owning men of Northern European ancestry.[8] Not surprisingly, then, possessive individualism has its source in a restrictive definition of the individual that *ex*cludes far more people than it *in*cludes. Historically and even today, many women of whatever colors and other marginalized peoples have been measured by and found lacking when compared to this exclusionary standard.

Self-enclosed individualism is premised on a binary model of identity formation, or what Kelly Oliver describes as a "conceptual framework in which identity is . . . formed and solidified through an oppositional logic that uses dualisms to justify either opposition and strife or awkward or artificial bridging mechanisms" (*Witnessing* 51). In this configuration, individualism sets up a hierarchical relationship between self and other, where the individual and society occupy mutually exclusive poles. This hierarchy presumes and reinforces a model of domination, scarcity, and separation in which intense competition leads to aggressiveness and fear: *my* growth requires *your* diminishment. Interactions between self and other are conflict-driven, and society is reduced to a collection of individuals motivated by greedy, insatiable self-interest. (It's survival of the fittest, so to speak.)

This hyper-individualism and the oppositional framework on which it rests prevent us from recognizing our interconnectedness with others and working together for social change. When we[9] live our lives according to this story of rugged individualism, we adopt a competitive model of success. Distancing ourselves from all that surrounds us, we become defensive, isolated, and alienated. We assume that success depends only on individual effort: Those people who do not succeed have only themselves to blame, and their failure has absolutely no impact on anyone but themselves. This highly individualistic story is deeply infused into mainstream U.S. culture and greatly inhibits social justice work. After all: if each individual is fully responsible for his or her own life, there is no need for collective action or systemic change. Just pull yourself up by your own bootstraps! Oliver makes a related point in her discussion of contemporary racism and sexism: "The individualism behind notions of formal equality and a color- and gender-blind society reduces social problems to personal sins on the part of whites and men and mental instability or physical defects on the part of people of color and women" (*Witnessing* 163). As Oliver's statement implies, the status-quo story of possessive individualism denies that color, economic status, gender, sexuality, and other human variables shape people's lives in different ways.[10]

Although it might sound paradoxical, hyper-individualism—with its apparent celebration of each person's unique traits—is based on a hidden assumption of sameness defined in extremely narrow terms. This restrictive individualism contains a monolithic definition of the individual in which all individuals are, in some very basic ways, identical and should be treated exactly the same. As powell notes, "individuality, even as it purports to take into account our distinctness, makes us all the same in fundamental ways. We are all rational, autonomous people and therefore we should all be treated the same" ("Disrupting Individualism" 9). This assumption of sameness and its denial of difference discredits affirmative action, reparations, or any other attempts to address long-standing historically based inequalities.

Like most U.S. Americans, the majority of my students have been seduced by these status-quo stories of rugged, self-enclosed individualism.[11] Defining themselves as fully independent and autonomous human beings, they have adopted and try to enact the competitive model of success I described above. When they perceive themselves according to this status-quo story, they cannot recognize their interconnections with or accountability for others. Nor can they perceive or understand the continuing significance of gender, 'race,' and the other social categories that measure and mark us. Instead, they are firm believers in rugged individualism and the related status-quo stories of meritocracy, equal opportunity, and fair treatment for all. These stories have persuaded many of my students (and of course numerous other U.S. citizens as well) that the United States is a free democratic country where everyone is treated the same, everyone has the same obstacles, and everyone has the same opportunities to prosper: In the United States, anything is possible and all doors will open to those who work hard. As Derald Wing Sue notes, these false stories (or what he describes as a series of interlocking "myths") function to erase the ways racism and other forms of oppression restrict some people's actions and greatly inhibit their success:

> The myth of meritocracy operates from the dictum that there is a strong relationship between ability, effort, and success. Those who are successful in life are more competent, capable, intelligent, and motivated. Those who fail to achieve in society are less capable, intelligent, and motivated. The myth of equal opportunity assumes that everyone encounters the same obstacles in life and that the playing field is a level one. Thus, everyone has an equal chance to succeed or enjoy the fruits of their labor. The myth of fair treatment equates equal treatment with fairness, whereas differential treatment is considered discriminatory or preferential. All three often act in unison to mask disparities and inequities and to allow actions that oppress groups that are not in the mainstream. (767)

This hyper-individualism and its false stories can make students callous and judgmental: Because they believe that each individual is the master of her or his own fate, fully responsible and accountable only to her- or himself, they blame the individual for his or her failure to succeed. "Oh sure," say some of my students, "racism and sexism *might have* occurred in the past, but thanks to the Civil Rights Movement and the Women's Movement, they have been solved. We're all equal now. There is no need for feminism, anti-racism, or those other social movements."

In addition to ignoring the systemic nature of social injustice, this dogma of meritocracy allows 'white'- and/or male-identified[12] individuals to deny the ways color, gender, and other markers of difference materially impact people's lives. As Harlon Dalton observes, "For a significant chunk, the inability to 'get' race, and to understand why it figures so prominently in the lives of most people of color, stems from a deep affliction—the curse of rugged individualism" (105).

forging "new" stories

The deeply ingrained status-quo stories of individualism, meritocracy, equal opportunity, and fair treatment for all have dominated our personal, professional, and socioeconomic lives for too long. With their emphasis on separation, division, and rigid classification, these status-quo stories have not served us well; and they are too limited to bring about radical transformation. Like Elana Dykewomon, I believe that "the power of a compelling story" can change the world (454). We need different stories about reality, stories that enable us to question the status quo and transform our existing situations. We need "new" stories informed by connectionist thinking and infused with the recognition of our interrelatedness. I mark the word "new" to indicate that stories of interconnectedness, though they might seem new to some people, are actually extremely old, for they reflect what Gregory Cajete refers to in my second epigraph to this chapter as "the ancient idea of relationship." Unlike the status-quo stories circulating in today's westernized cultures, these ancient-yet-new stories of interconnectedness can stretch our imaginations and in other ways empower us to change. These stories teach us that we are interrelated and interdependent—on multiple levels and in multiple ways. To mention only a few of the many ways we are interconnected, I suggest the following:

We Are Interconnected Economically. The growth of multi- and transnational companies, "the creation of a global fiber-optic network" (Friedman 10), recent developments in software, new forms of outsourcing, off-shoring, and other changes in the ways companies and individuals do business has lead to the development of "a single global network" (Friedman 8). Economic changes in one location often have a domino-like effect on local, national, and even global communities.[13]

We Are Interconnected Ecologically. We are rooted in, sustained by, and interdependent with the environment.[14] As the Chagga people of Tanzania teach, "the universe is a web of one interconnected, interrelated, and interdependent whole . . . Stones and mountains, rivers and lakes, clouds and rain, are all *alive* in their intrinsic meanings and in their active partnership to people and everything else" (Mosha 213, his emphasis). Not only are we intimately embedded within the natural world, but this world reinforces our interconnections with each other.

In fact, it has been "roughly calculated that every breath of air we breathe contains a few atoms that have been breathed by every person in the history of the planet, from Socrates to Genghis Khan to Einstein to Hitler, as well as all the billions of unknowns" (Hayward 21).

We Are Interconnected Linguistically. Language itself is a material reality that ensures our interconnections with others. The words we use come to us already infused with meaning and value. Although we appropriate and shape these words to fulfill our own intentions, these intentions are never purely our own. As Mikhail Bakhtin suggests, we are born into consciousness through the words of others: "All that touches me comes to my consciousness—beginning with my name—from the outside world, passing through the mouths of others . . . with their intonation, their affective tonality, and their values."[15]

We Are Interconnected Socially. Even the labels we use are relational. Terms such as "male"/"female," "heterosexual"/"homosexual," "parent"/"child," "self"/"other," and "black"/"white" come into being relationally and have meaning only in dialogue with each other. As powell reminds us, "one's own sense of identity is inextricably entwined with, and dependent upon, the identity of 'others' " ("Multiracial Self " 1498).

We Are Interconnected Spiritually. According to some Indigenous and non-Indigenous teachings, there exists a cosmic, constantly changing spirit or force that embodies itself as diverse material and nonmaterial forms.[16] As Cajete explains, for many Tribal peoples, "[e]very act, element, plant, animal, and natural process is considered to have a moving spirit with which humans continually interact" (*Native Science* 69). Some non-Tribal people have held similar beliefs. According to that well-known nineteenth-century Transcendentalist, Ralph Waldo Emerson, "[n]ature is not fixed but fluid. Spirit alters, moulds, makes it" (*Nature* 48). For contemporary theorist Gloria Anzaldúa as well, "[s]pirit exists in everything; therefore God, the divine, is in everything—in . . . rapists as well as victims; it's in the tree, the swamp, the sea" (*Interviews/Entrevistas* 100).

But my point here is not to prove that this interconnectedness "really" exists, defining "really" in some scientistic fashion. Instead, I am interested in using interconnectivity as a framework on which to develop transformative theories, pedagogies, and social action. One of my primary goals as an educator is to awaken in my students a sense of our radical interconnectedness, for I am convinced that this awareness can play a crucial role in working toward social justice. At the very least, I offer relational worldviews as alternatives to the highly celebrated belief in an entirely independent "American" self. Seduced by stories of rugged individualism, the majority of my students (especially at the undergraduate level) do not believe that their actions have consequences for anyone but themselves. Nor can they recognize that what affects others—*all* others, no matter how separate we seem to be—ultimately affects them as well.[17] And so, I posit radical interconnectedness as my theoretical and pedagogical framework and develop classroom strategies designed to challenge students' perceptions. Framing my pedagogy in this way serves several purposes.

First, this framework enables me to redefine individual and communal identities and thus offer an alternative to possessive individualism. Locating each individual within a larger, holistic context, I maintain that human beings are not entirely self-enclosed. We are indeed distinct entities, but we also share permeable boundaries. As Anzaldúa explains, "[t]he self does not stop with just you, with your body . . . [T]he self can penetrate other things and they penetrate you" (*Interviews/Entrevistas* 162). Our permeable selves extend outward . . . meeting, touching, entering into exchange with other subjects (human and nonhuman alike). Significantly, this outward movement is not an imperialistic appropriation where an egocentric subject grows larger by extending its boundaries to incorporate or annihilate every object in its path. It is, rather, a mutual encounter between subjects.[18] This definition invites students to shift perspectives and adopt a much broader—though inevitably partial—point of view.

I use this concept of permeable selfhood to replace the stories of solipsistic individualism with relational forms of individualism that value both personal and collective integrity and self-respect. In this relational model of human nature, each person's self-definition and self-growth occur in the context of and in dialogue with other equally important individuals. (I define "individual" broadly to include both human and nonhuman life.) We are both distinct individuals and integral parts of a series of larger wholes. Significantly, these "larger wholes" do not imply homogeneity or sameness. Living systems theorists offer a useful analogy for this both/and approach in their concept of the holon. Coined by Arthur Koestler from the Greek word *holos* (which means whole), a holon is, simultaneously, an *autonomous* system and a vital part of a *larger* system. As Joanna Macy explains, "All living systems—be they organic like a cell or human body, or supra-organic like a society or ecosystem—are *holons*. That means they have a dual nature: They are both wholes in themselves and, simultaneously, integral parts of larger wholes" ("Living Systems," her emphasis).[19] We—each of us, distinct human beings—are both self-contained and collective, both open and closed. We do have boundaries, but these boundaries are porous, allowing exchange with our external environment. The divisions between self and other, between individual and world, still exist; however, these divisions are not nearly as rigid and inflexible as that status-quo story of rugged individualism would lead us to believe. Social activist Fran Peavey offers a useful analogy for human interconnectedness:

> Human beings are a lot like crabgrass. Each blade of crabgrass reaches up to the sun, appearing to be a plant all by itself. But when you try to pull it up, you discover that all the blades of crabgrass in a particular piece of lawn share the same roots and the same nourishment system. Those of us brought up in the Western tradition are taught to think of ourselves as separate and distinct creatures with distinct personalities and independent nourishment systems. . . . Human beings may appear to be separate, but our connections are deep; we are inseparable. Pull on any part of our human family and we all feel the strain. (13)

I appreciate this description because it acknowledges our apparent separateness while insisting on our deep interdependence.

Second, and closely related, using interconnectivity as my pedagogical framework enables me to develop alternatives to possessive individualism's binary-oppositional models of identity formation. Generally, identity formation functions through exclusion: we define who and what we are by excluding who and what we are not. I am especially interested in developing *inclusionary* identities, and I believe that interconnectivity itself might serve as a springboard enabling me to imagine and thus create nonbinary identities at both personal and communal levels. Anzaldúa makes a similar point. As she explains in a 1991 interview, she believes that we share an interconnectedness that could serve as "an unvoiced category of identity, a common factor in all life forms" (*Interviews/Entrevistas* 164). This "common factor" goes beyond—*but does not ignore*—social categories based on gender, 'race,' or other systems of difference; because it is "wider than any social position or racial label" ("now let us shift" 558), it can include them. Indeed, this shared identity is wider than anything in "human nature." As Anzaldúa explains, each person's identity "has roots you share with all people and other beings—spirit, feeling, and body comprise a greater identity category. The body is rooted in the earth, la tierra itself. You meet ensoulment in trees, in woods, in streams" ("now let us shift" 560).

Positing interconnectedness as itself a shared trait, I can replace binary self/other systems of difference with a more expansive, relational approach to identity. This approach does not deny the validity of each person's specific experiences, beliefs, and desires. Instead, it enables me to reposition individual differences within a larger, holistic context—a context premised on the possibility of commonalities. Let me emphasize: This repositioning does not negate individual differences, for "commonalities" and "sameness" are not synonymous. As I define the term, commonalities contain—without erasing—differences. Commonalities occur as we negotiate among sameness, similarity, and difference—providing a synergistic mixture brewed from all three. I refer to this mix as *synergistic* to underscore its unpredictable, creative dimensions. As I illustrate in more detail in the following chapter, commonalities are points of connection; they offer pathways into unexpected, complex interactions with others. When we explore commonalities, differences do not wither away; however, they do become less divisive. We don't need to break the world into rigid categories and hide behind masks of sameness which demand that we define ourselves in direct (binary) opposition to others.[20] We can trust that, despite the many differences among us, we are all interconnected.

Using interconnectivity as my pedagogical framework serves a third purpose by enabling me to motivate my students. In my classrooms, the recognition of our interconnectedness can fuel students' desire to work for social change. Because self and other are irrevocably, utterly, *intimately* interrelated, what affects you—no matter how distant, how separate, how different (from me) you seem to be—affects me as well. This perception of interdependence generates a matrix of reciprocity, mutual accountability, and respect. Students recognize that the events and belief systems impacting our sisters and brothers across the street, across the state, across the globe, have a concrete effect on us all. We all rise or sink together. David Loy makes a similar point:

> This world is not a collection of objects but a community of subjects, a web of interacting processes. Our "interpermeation" means [that] we cannot avoid responsibility for each other. This is true not only for the residents of lower Manhattan, many of whom worked together in response to the World Trade Center catastrophe, but for all people in the world, however hate-filled and deluded they may be . . . including even the terrorists who did these horrific acts, and all those who supported them. (*Great Awakening* 108, his ellipses)

As I demonstrate in the following chapters, I design my course syllabi and class discussions in ways that invite students to recognize this "interpermeation." I do not require or demand that they accept my belief in our interconnectedness. As my use of the word "invite" suggests, I simply encourage them to consider this possibility and explore its consequences.[21] I reinforce this invitation by attempting to enact an ethics of openness, reciprocity, and exchange based on my belief in our interrelatedness.

toward an ethics of openness, reciprocity, and exchange

This ethics begins with how I perceive my students. I assume that they—like all people—have intellectual/intuitional agency: they can think for themselves and, through careful self-reflection, can choose to alter their beliefs, worldviews, self-definitions, and actions. I don't know for sure whether my students will change, but they have the potential to do so. Like other people, my students have "blank spots": ignorances, resistance to knowing certain things, and so forth—or what Anzaldúa refers to as "desconocimientos" or "willful unawareness"[22]—but they generally try to do their best. I believe that when I respect my students and their potential, they will live up to my expectations.

I share these perceptions with my students by including extensive information, or what I call "Dialogue: Some of My Presuppositions," in each syllabus (appendix 1). I revise this information at the end of each semester, altering it to address my most recent classroom experiences. Currently, this "Dialogue" includes six presuppositions, designed both to share my perspective with students and to anticipate possible objections to some of the more contentious issues we will explore:

1. *Social injustice exists. People are not treated equitably. We live in an unjust society and an unfair world. Although "liberty," "equality," and "democracy" are radical ideas with great promise, they have not yet been fulfilled. Oppression (racism, classism, sexism, etc.) exists on multiple seen and unseen levels.* I include this first point, insisting on the existence of social injustice, because I don't want to spend class time debating whether or not sexism, racism, and other forms of injustice "really" exist. As my reference to liberty, equality, and democracy indicate, I do not focus exclusively on the negative dimensions of U.S. culture; by alluding to some of this nation's most highly celebrated (though too rarely seen) ideals, I assure my students that I am optimistic about our country's potential.

Oppression is not inevitable; we can envision alternate possibilities. I also use this reference in order to anticipate possible objections to what some students have characterized as an "anti-American" bias. I want students to understand that I am not biased against the United States. Rather, I take seriously this country's rhetoric of freedom, democracy, and justice for all.[23]

2. *Our educations have been biased. The eurocentric educational systems, media outlets, and other institutions omit and distort information about our own groups and those of others. These hidden mechanisms sustain oppression, including an often invisible and normative 'white' supremacy. Not surprisingly, we all have "blank spots," desconocimientos (Anzaldúa), and so forth.* With this emphasis on education, I challenge my students' intellectual individualism. While in many ways they can be autonomous thinkers, it's crucial that they recognize and explore the ways their beliefs have been subtly shaped by the dominating culture. This second point also enables me to set the stage for later discussions of complicity and accountability.

3. *Blame is not useful, but accountability is. It is nonproductive to blame ourselves and/or others for the misinformation we have learned in the past or for ways we have benefited and continue benefiting from these unjust social systems. However, once we have been exposed to more accurate information, we are accountable! We should work to do something with this information—perhaps by working towards a more just future.* I developed this third point especially for my 'white'-raced students and wealthy students of any color. Each semester, at least one student (usually 'white'-raced), will begin feeling guilty about racism, enslavement, and other forms of racial injustice. I want to prevent this solipsistic guilt because it is counterproductive. In addition to shifting class attention to these guilt-ridden students, it paralyzes them.[24] If necessary during class discussions, I refer back to this point.

4. *"We are related to all that lives.": We are interconnected and interdependent in multiple ways, including economically, ecologically, linguistically, socially, spiritually.* I use this point to introduce the concept of interconnectivity and to offer an alternative to the binary thinking students generally employ.

5. *Categories and labels shape our perception. Categories and labels, although often necessary and sometimes useful, can prevent us from recognizing our interconnectedness with others. Categories can (a) distort our perceptions; (b) create arbitrary divisions among us; (c) support an oppositional "us-against-them" mentality that prevents us from recognizing potential commonalities; and (d) reinforce the unjust status quo. Relatedly, identity categories based on inflexible labels establish and police boundaries—boundaries that shut us in with those we've*

deemed "like" "us" and boundaries that shut us out from those whom we assume to be different. Because identity-related issues play such a large role in the courses I teach, we regularly discuss categories and labels. However, I want to do so without further reinforcing them in students' thinking. With this presupposition, I begin calling the categories into question even as I introduce their importance.

6. *People have a basic goodness. People (the authors, the groups we study, and class members) generally endeavor to do the best they can. We will all make mistakes, despite our best intentions. The point is to learn from our errors. In order to learn from our errors, we must be willing to listen and to speak (preferably, in this order!).* I borrow this concept of basic goodness from Buddhist teachings and use it to acknowledge the fact that mistakes are bound to occur. No one is perfect. It sounds so obvious, but in class discussions (especially in women's studies courses where students so often have firmly held, highly politicized beliefs!) students can become moralistic and judgmental. We can feel so righteous in our concerns about social injustice that we trample over those who have different views. Yet this self-righteous, reactionary attitude is quite different from the type of inclusionary classroom I want to create. By reminding students that it can be better to listen before we speak out, I encourage them to be more self-reflective and less reactionary.[25]

One of the most effective ways in which I demonstrate respect for my students' intellectual agency is by incorporating discussion questions into each course.[26] As I define the term, discussion questions are questions students have about one or more of the week's assigned readings that they would like us to explore in class. Discussion questions require deep engagement with the material; they must be able to generate sustained conversation. I don't give credit for easily answered yes/no questions or for questions that simply request the definition of a term, historical information, or other easily researched information. (For instance, questions such as "What does 'transgender' mean?" or "What is a mulatto?" or "When did the Harlem Renaissance occur?") A discussion question can deal specifically with one text or it can be a bit broader and engage several assigned readings. In graduate courses, I require that students email me their discussion question by a specified date and time (usually thirty-six to forty-eight hours before the scheduled class meets). In undergraduate courses, discussion questions are due at the beginning of the class. I take these discussion questions very seriously and use them to organize and guide class conversations. Discussion questions serve at least three additional purposes as well: First, they offer opportunities for students to reflect more deeply on the assigned readings and, through this reflection, to deepen their learning; second, they enhance student accountability and give students more control over our time together; and third, they allow me to assess student interests and learning.[27]

As I interact with my students I assume that we share some commonalities (not identical traits!), but I do not presume to know precisely what these commonalities might be. In classroom discussions I make space for what Anzaldúa describes as "an unmapped common ground" ("now let us shift" 570). Significantly, this common ground is "unmapped" and resists being defined in narrow terms. I, for one, do not have a precise definition. At the very least, though, I assume that this common ground includes what Evelyn Alsutany describes as "the desire to belong in this world, to be understood" (107). As I interact with my students, I believe that they, too, want to be understood. Sometimes, I search for points of connection and posit commonalities that my students are free to accept, adopt, revise, or reject. I watch as these possible commonalities develop or fade through class discussions, writing assignments, and other interactions.

Closely related, my ethics of openness, reciprocity, and exchange entails that I acknowledge each student's "complex personhood" (Cervenak et al. 352). Each person I encounter has a specific, highly intricate history, an upbringing and life experiences that I cannot fully know. I don't know all the forces that shaped my students and, at best, I can only partially ascertain their intentions and desires. I enter each classroom with the reminder that my understanding is always somewhat inadequate and incomplete. If, as I listen, I acknowledge both an unmapped common ground and each student's complex personhood, I can remain open-minded. I don't want to be trapped by the stereotypes often inhibiting the ability to listen fully, with open heart and mind. You know—those culturally sanctioned, monolithic categories that invite us to look at a person, label her, and assume that we *know* her position, motivations, values, and beliefs: We *know* her . . . because of the identity groups to which she seems to belong, because of her previous comments, or because of other such signs. I remind myself not to fall into stereotypical thinking: Maybe I *do* know what she'll say, but can I be certain that I fully understand what she means? Do I know the intentions and desires behind her words? When I'm convinced that I entirely know a student, I close my mind and shut down. I do not fully engage with his or her words. (I've heard them already, I've heard them many times.)

My ethics of openness, reciprocity, and exchange challenges me to go beyond my academic training and take new kinds of risks. Like others trained in the academy, I have honed my debate skills, learned to think on my feet in ways that interrupt my ability to listen fully. When I use these carefully developed skills in my classrooms, I can't listen very well. Rather than simply *open* myself to students' words, I'm busy anticipating my next response. This openness is especially challenging when students voice sexist, 'white'-supremacist, ethnocentric, and/or other damaging views. I want to challenge them immediately and point out their statements' enormous errors! However, I have learned that I can be more effective (for all students) when I slow down and take a less adversarial approach. As I demonstrate in the following chapters, this approach demands temporary detachment from self, coupled with a nonjudgmental attitude. This combination of detachment and nonjudgment prevents me from taking the words I hear personally and reacting automatically. Instead, to be fully present with students as they speak, I disengage from my personal feelings and self-image. Opening myself fully, I listen. I observe each person as s/he speaks. I read my students' body language: the hunched shoulders, the clenched jaw, the averted eyes. I listen with my body, allowing the feelings and words to penetrate me deeply.

This temporary self-detachment enables me to expand my awareness, and I listen for students' silences, erasures, omissions. As Anzaldúa suggests in her discussion of nepantlera conflict resolution, "By attending to what the other is *not* saying, what she's *not* doing, what isn't happening, and by looking for the opposite, unacknowledged emotion—the opposite of anger is fear, of self-righteousness is guilt, of hate is love—you attempt to see through the other's situation to her underlying unconscious desire" ("now let us shift" 567, my emphasis). But of course, I can't know for sure what these underlying desires might be and I could misinterpret the words actually spoken. (After all, words have multiple meanings.) As Cervenak and her coauthors note, "misunderstandings and miscommunications occur because we assume we understand the meaning of everything spoken, when we often do not" (346). So, I ask questions to test my assumptions and clarify what I believe students have said. I paraphrase their words and ask them if I have paraphrased correctly. When necessary, I ask for additional clarification.[28]

Conflict is perhaps inevitable when we bring social-justice issues into our teaching, but I try to avoid overly simplistic two-sided debates. In my syllabi and introductory lectures, I acknowledge the possibility of conflict and disagreement. "You won't always agree with me or with each other," I explain, "and at times perhaps you'll disagree quite strongly.

When faced with conflict many of us want to go into avoidance mode: ignore it, deny it, hide from it, stop talking, stop attending class, or in other ways withdraw. Some of us might try to solve the disagreement too quickly, without fully understanding the needs, desires, beliefs, and worldviews that triggered and drive it. And still others might want to just celebrate our diverse opinions, agree to disagree about a point of contention, and move on." I warn them that these reactions are not productive. When we handle conflict through avoidance, a fix-it mentality, or a superficial "let's-just-agree-to-disagree" attitude, we remain entrenched in our already-existing beliefs. Where is the room for change and growth? Agreement—if defined as sharing the exact same view—is neither necessary nor productive. We need commonalities—not identical opinions and perspectives!

There are lessons in each conflict, I assure my students, especially when we can approach the conflict with a degree of detachment, self-reflection, and the desire to learn and be transformed. Conflict, though painful, can change us in life-enhancing ways. As Anzaldúa notes, "Conflict with its fiery nature can trigger transformation depending on how we respond to it. Often delving deeply into instead of fleeing from it can bring an understanding (conocimiento) that will turn things around" ("(Un)natural bridges" 4). I encourage students to explore the conflict self-reflectively, reminding ourselves that if we listen with open hearts and open minds we might, together, forge new visions.

I am still learning how to enact this ethics of openness, reciprocity, and exchange in my classrooms, and I probably fail more often than I succeed. But like conflict, failure can facilitate transformation, and failure's lessons can lead to success. Indeed, many of the strategies I discuss in *Teaching Transformation* had their source in excruciatingly painful classroom investigations, conflicts, and failures. And so, I hold tightly to my belief in our radical interrelatedness. With every classroom encounter, I make the conscious decision to act on our interconnectedness—even when these connections are difficult or impossible to perceive. Using imagination and mindful attention, I posit, explore, and in other ways try to enact interrelatedness. When failure occurs, I remind myself that we are always in process, always open to further change.

shifting perception, changing stories

Language, belief, perception, and action are intimately interrelated, shaping our reality in complicated ways.[29] The stories we tell ourselves, the stories we learn from our cultures (our families, schooling, religion, friends, the media, the books we read, and so forth) influence our beliefs, our perceptions, and our actions. As Stuart Hall asserts, "How we 'see' ourselves and our social relations *matters*, because it enters into and informs our actions and practices" ("Racist Ideologies" 272, his emphasis). What we believe about ourselves, other people, and the world affects our perceptions, and these perceptions affect how we act; our actions move out into the world, affecting other people's beliefs, perceptions, and actions.[30] Our actions shape the stories we tell, the stories others tell about us, the ways they perceive us, the ways we perceive ourselves. It's an interlocking web with no clear-cut beginnings or endings.[31]

All too often, however, we assume that our perceptions and beliefs accurately reflect the entire truth about reality and ourselves. Like the status-quo stories I've described in this chapter, this assumption narrows, limits, restricts our worldview and inhibits our actions. Linda Myers makes a similar point: "The reality we perceive is shaped by an underlying system of beliefs often implicit, assumed, or unquestioned, which serves as a self-fulfilling, self-prophetic organizer of experience" (*Understanding an Afrocentric View* 31). All too often, these underlying beliefs are so informed by status-quo stories that they shut down imaginative possibilities and prevent us from recognizing our responsibility as co-creators.

I reject the status-quo stories explored in this chapter, and I invite my students to do so as well. I want my students to peel away these status-quo stories, expose the implicit beliefs they represent, analyze the stories' impact on their lives and the world, and create new, more progressive stories.

To return to my opening anecdote, these normalizing concepts—self-enclosed individualism, binary-oppositional thinking, and the inflexible permanence of 'race' and other identity categories—function in our collective psyche like computers on the radiator. They are so deeply embedded in the dominant-cultural stories we are exposed to every day, that we generally assume they are unchanging aspects of our world. "It's *always* been that way." Yet these concepts damage us.[32] Maybe they *are* permanent and unchanging, but maybe they're not. As this chapter has demonstrated, I choose to believe that status-quo stories can—and must!—be changed. In my classrooms, in this book, in other areas of my life, I am forging new (or, rather, extremely old yet too often forgotten) stories about interrelatedness—stories that can assist us in making social change.

CHAPTER TWO

Forging Commonalities: Relational Patterns of Reading and Teaching

Like an individual, America can be whole only by going back to its roots—all of them. My premise is this: the Native American story—and the holistic mode of thought it embodies—springs from the original root in our homeland. The story is designed to move among the strands of life's web both within the individual and within the community, to restore balance and harmony. Its ancient ways offer a helpful pattern in making new connections among our different people and academic disciplines.

Marilou Awiatka[1]

One rainy November morning a few years ago as I was driving my six-year-old to school, she leaned toward me, stared at my ears, and stated, "Those are war earrings, Mama." I was startled. Given the proximity to Thanksgiving I had a strong hunch that she was associating my arrow-shaped earrings with "Indians"—defined simplistically as "warriors." We were late for school and I couldn't immediately address my concern, but subsequent conversations confirmed my suspicions: Six years of home training (children's books by Simon Ortiz, Luci Tapahonso, Joy Harjo, and others; age-appropriate critiques of Disney's *Pocahontas*, Columbus Day, and so forth) had apparently been wiped out by a single teaching unit in her first-grade classroom: "Indians" had become a monolithic stereotype relegated fully to the past, a past that has nothing to do with "us" here today. I was reminded, once again, of the power—not always positive—of formal education, and why I am committed to transformational multiculturalism. Unlike the celebratory melting-pot versions of multiculturalism my daughter was experiencing at school, transformational multiculturalism explodes such stereotypes and exposes the 'white' supremacist narrative from which they emerge and on which they rely.

Transformational multiculturalism employs relational reading and teaching strategies, creating what Awiatka describes in my epigraph as "a helpful pattern in making new connections among our different people and academic disciplines." In this chapter, I discuss some of the relational teaching techniques I have developed as I try to forge these new connections.

relational reading, relational teaching

Relational patterns—structured by syllabi, text selection, class discussion, and assignments—begin with commonalities and take multiple inter-locking directions. This relational approach facilitates what Françoise Lionnet describes as "nonhierarchical connections" among differently situated cultural and historical traditions: "Instead of living within the bounds created by a linear view of history and society, we become free to interact on an equal footing with all the traditions that determine our present predicament. On a textual level, we can choose authors across time and space and read them together for new insights" (7–8). To return to my opening example, my daughter's history lesson about "The First Thanksgiving Day" illustrates the linear historical view Lionnet critiques: the encounter between "Indians" and "Pilgrims" occurred during a time in the distant past, among two groups of people who no longer exist. The only possible connection between these long-ago people and "our present predicament" is the annual celebration of this supposedly first encounter. For my daughter and her classmates, this commemorative event authorizes and reaffirms the dominant-cultural belief that present-day America represents the indigenous inhabitants' willing transfer of ownership.

Many of us have been trained to regard history in a similar fashion: as a linear timeline that clearly separates the present from the past and "us" from "them." When viewed from this perspective, history is reduced to "a series of fixed points on an abstract historical continuum" (Ratcliffe 109) that prevents us from recognizing how situations and events that occurred in the past continue to influence our existence today. Locating the past entirely in a time before ourselves, we separate ourselves from historical injustices and so cannot recognize that the land theft, human bondage, and other forms of conquest that began hundreds of years ago continue informing the present. Not surprisingly, then, this fragmented, linear perspective establishes boundaries that deny accountability.

In our classrooms, relational patterns of reading can challenge students to recognize that the past is not elsewhere; it is with us today. Our worldviews, our understanding of what it means to be "American," our interactions with those we define as "other," are shaped by and shot through with earlier events, attitudes, values, and beliefs. As Ratcliffe notes, the past is "a series of inscriptions in discourse and on our material bodies, inscriptions that continually circle through our present and form our identities, inscriptions that will control us if we do not acknowledge them" (109). Relational patterns invite readers to acknowledge and explore these circulating inscriptions and thus begin recognizing history's continual impact on contemporary U.S. life. This recognition is crucial for progressive social change. In the classroom, relational teaching combines cultural critique with self-reflection. By so doing, it enables students to examine both their own often unconscious presuppositions and the various sources and effects of these previously unexamined beliefs.

Although I focus here on cross-cultural relational patterns, I am not by any means denying the importance and value of cultural/ethnic-specific explorations. Because there are so many differences within each culturally specific tradition, this relational approach can be useful in culturally specific courses as well. We need both culturally-specific and multicultural courses; yet especially in smaller, budget-constricted schools, we cannot always have the number of ethnic- and cultural-specific courses we need. While I respect (and share with my students) the view expressed by Kelly Morgan that "the study of American Indian literature should be separate from the study of American literature" (30),[2] I consider American Indian literatures to be a vital part of the U.S. literary canon. Like Joy Harjo, I believe that "[t]he literature of the aboriginal people of North America defines America" ("Introduction" 31).

I incorporate Native texts, perspectives, and worldviews into every course I teach, and I do so for two reasons. First, I believe that it is crucial to challenge misconceptions about First Nations peoples. The majority of my students have very little knowledge of Native American histories, cultures, or literatures. In fact, their views are almost as simplistic as those of my six-year-old and her first-grade classmates. Like most people trained in mainstream U.S. educational systems, my students regard Indians (when they think of them at all) in highly romanticized, ahistorical terms: as primitive, essentially natural peoples (all wearing feathers, living in teepees, running around with bows and arrows, and so forth). They know little—if anything—about issues of Native sovereignty, the 1830

Indian Removal Act and the subsequent Trail of Tears, the Indian Boarding Schools, the 1887 Dawes Act, the 1924 Indian Citizenship Act, or other aspects of the ongoing history of conquest in this nation.[3] This historical amnesia and other forms of "desconocimientos"[4]—or willed ignorance—keeps people falsely innocent and naive, and prevents them from recognizing the connections between U.S. colonialism in the past and the contemporary forms U.S. imperialism takes both at home and abroad. Second, I incorporate Indigenous texts, perspectives, and worldviews into my courses because I believe that the "holistic mode[s] of thought" (Awiatka 155) they generally (but not always) embody offer alternative perspectives that can transform student thinking. As Ladislaus M. Semali and Joe L. Kincheloe explain, "[w]hen Western [sic] epistemologies are viewed in light of indigenous perspectives, Western ways of seeing, Western education cannot remain the same" (32).

I describe transformational multiculturalism's teaching patterns as *relational* to underscore what I believe to be a vital but too often ignored insight for contemporary life: We are all interconnected. As I asserted in the previous chapter, we are interrelated and interdependent—on multiple levels and in multiple ways: economically, ecologically, linguistically, socially, and spiritually . . . to mention only a few of the many ways we are interconnected. We are interlinked in every way we can possibly imagine, as well as in ways that perhaps we cannot yet fathom.

forging commonalities

Relational teaching begins with commonalities. I define commonalities neither as sameness nor as *identical* experiences, histories, ideas, beliefs, or traits. Instead, commonalities indicate complex points of connection that enable us to negotiate among sameness, similarity, and difference. The trick (and it can be tricky!) is to shape discussions of commonalities in ways that neither invite nor permit students to assume that their experiences, histories, ideas, or traits are identical with those of others. Such assumptions of sameness facilitate appropriation. In my classrooms, commonalities have taken many forms, including (but not limited to) region, genre, topic, or theme. We take a similarity like "creation" or "the land," and trace its journey through different texts, noting the many ways it changes and investigating the sociocultural, historical, power-laden implications of these alterations. The commonalities I select are based both on each specific course and on student interests, backgrounds, and concerns. To be sure, it's important to educate students about the many differences among us—and believe me: I do! However, I find it most effective to begin explorations of difference by focusing on commonalities.

Beginning with commonalities serves at least three interrelated purposes. First, it offers effective entry points into class discussions. Because I try to develop commonalities that will resonate with my students, I select topics that they have thought about, experienced, or are in other ways at least somewhat familiar with. This approach makes students more willing to engage with the material; confident that they can contribute to our discussions, they are often eager to speak out. So, for example, when I taught composition at an open-admissions public university attended primarily by first-generation, working-class college students who equated a college degree with economic advancement, I found it effective to begin with class-based commonalities such as "the American Dream." I divided the dream into four parts—images of America, definitions of success, formulas for achieving success, and views of individual agency—which we explored through a variety of historical and contemporary texts written from diverse perspectives.[5] Focusing first on class-based issues (although in the classroom I did not generally refer to them as such), allowed me to engage students in dialogue.

Second, beginning with commonalities enables me to encourage potentially transformational self-reflection by inviting students to (re)examine their own presuppositions and worldviews. This self-reflection can be difficult and uncomfortable, but like Louis Owens, I believe that "[i]t is our responsibility, as teachers and writers, to make sure that our texts and our classrooms are not 'safe' spaces from which a reader or student may return unchanged or unthreatened. . . . It is our job . . . to make people listen well, to disrupt the discourse of dominance, to challenge and discomfit the reader, to ultimately startle that reader into real knowledge" (46–47). As Owens implies, this transformational process entails interrupting and challenging the "discourse of dominance"—the 'white' supremacist framework that naturalizes and thus reinforces our unjust status quo. Focusing on commonalities enables me to expose this framework in subtle yet potent ways. I denaturalize the dominating norm by putting it under investigation—in dialogue with other (nondominant) cultural narratives. Such exposure and denaturalization are crucial to transformational multiculturalism. Stripping away the unspoken authority this framework has automatically acquired (Sandoval 93–95; Lakoff 50–54), this dialogue invites students to perceive their own worldviews and assumed histories from different perspectives, thus startling them into new points of view.

Third, exploring commonalities prevents students from equating difference with deviation. Those of us educated in western-dominant cultural systems have been trained to define differences oppositionally—as deviations from what Audre Lorde terms the *"mythical norm, which . . . [i]n america [sic] . . . is usually defined as white, thin, male, young, heterosexual, christian, and financially secure"* (*Sister Outsider* 116, her emphasis)—and to regard these differences as shameful marks of inferiority. Driven by our fear of difference-as-deviation, we hide our differences beneath a facade of sameness and erect rigid boundaries between self and other. However, an exclusive investigation of the other(s)—be it other(s) defined by ability/health, class, color, culture, ethnicity/'race,' gender, religion, or sexuality—inadvertently normalizes this unexamined "mythical norm" and, by extension, implies the abnormality of the text(s) and/or group(s) under examination. As Paula Rothenberg asserts, "[w]here white, male, middle-class, European, heterosexuality provides the standard of and criteria for rationality and morality, difference is always perceived as deviant and deficient" (43). Investigating the dominant cultural narrative (a narrative that, in various ways, my students have unthinkingly consumed through many years of formal education) along with alternative narratives and worldviews makes it difficult, if not impossible, for them to dismiss the latter as "abnormal." Difference can no longer function as rigid division between "us" and "them."

To illustrate one form this relational approach can take, I now want to explain how I begin my survey of Colonial/U.S. American literature course. I focus on this course because it has played the largest role in teaching me how to teach relationally.

"in the beginning . . . "

Like many Americanists, I begin my U.S. survey course with Native creation texts, which we read in dialogue with Judeo-Christian creation stories. I have used the *Heath* anthology and for the first reading assignment require students to read the Zuni account, "Talk Concerning the First Beginning," along with a handout: the first three chapters in Genesis (King James Version). Here, the overarching commonality is "creation," a topic with which, in various forms, most of my students—who frequently identify as "Born-Again Christians"—are familiar. Yet my students are generally reluctant to recognize *any* commonalities between the two texts. As the following pages demonstrate, I have learned to use students' resistance to acknowledging commonalities as an opportunity to encourage increased self-reflection. During class discussions I validate students' objections and encourage them to explore the implications of their concerns. By so doing, I invite them to investigate and reflect on their own previously unexamined worldviews and the dominant-cultural framework that supports them. Like the majority of my students, I was raised in a fundamentalist "Born-Again Christian" home; I mention this fact because I believe that this commonality between myself and my students gives me insight into their objections and assists me in addressing their fears and concerns in respectful yet challenging ways. This respect is crucial, for it transforms students' rigid, suspicious resistance into a willingness to speak up, self-reflect, and entertain alternative perspectives.

The first objection occurs almost immediately, when I refer to both the Genesis chapters and "Talk Concerning the First Beginning" as "myths." Students experience an almost visceral reaction and challenge my description of Genesis as a myth. Rather than directly answer, I ask them to define "myth." Through class discussion, we[6] learn that they associate "myths" with "falsehoods" and "lies" or with inaccurate accounts of reality held by childlike, "primitive" peoples. They assume that by describing Genesis as a myth I'm implying that the Genesis story (and therefore the Bible as a whole) is untrue and that those who believe in its veracity are superstitious, ignorant, and wrong. Although at this point I am often tempted to defend myself by clarifying my position, I don't. Instead, I move the conversation outward by noting that no one seemed to object when I described "Talk Concerning the First Beginning" as a myth. Why not? I explain that I interpreted their silence as acceptance: they view the Zuni narrative as a nonfactual, entirely false story and, by extension, they view the Zunis themselves as a primitive, ignorant people whose lack of scientific knowledge prevents them from knowing "the Truth." Generally, this interpretation of their silence surprises students and compels them to self-reflect. As we work through the implications of my comments, we discover their unconsciously held assumptions about the superiority of their Judeo-Christian tradition and, more generally, western culture as a whole.

I then share my own quite different perspective: Unlike my students, I do not define myths as falsehoods, lies, or unscientific accounts held by so-called primitive peoples. Instead, I view myths as sophisticated, highly complex philosophical systems that express and offer insight into people's worldviews—their cosmologies, metaphysics, values, ethics, and so forth. I offer an overview of the extensive scholarship in Native American Studies that supports my perspective. This approach enables me to validate student concerns while inviting them to examine and reconsider their previous assumptions. Whereas I don't require or even necessarily expect my students to relinquish their assumptions, I do want them to recognize the ways these assumptions can influence their inter-pretations of the texts. And by placing both the Genesis chapters and "Talk" in this redefined context, I encourage my students to view *both* readings with respect.

Once we have worked through students' objections to my use of the word *myth*, I offer them a framework for class discussion. I explain my goals for this reading unit: I assign these texts so that we can explore the worldviews of the peoples involved in first contact/conquest. By reading the excerpts from Genesis and "Talk Concerning the First Beginning" in conversation, we'll attempt to obtain an understanding (partial and fragmentary though it will be) of the beliefs and values motivating the European explorer-conquerors and the peoples they encountered.[7] I carefully distinguish between stereotypes and generalizations, and caution students that we cannot assume to be talking about *all* Christians or *all* Native peoples. Such assumptions play into stereotyping. As Ian Haney-López explains, "A generalization abstracts on the basis of large numbers, with no pretense of specificity, while stereotypes ascribe unvarying characteristics to every individual within a group. Unlike those who engage in stereotyping, those who generalize remain acutely aware of the partial nature of their statements" ("Community Ties" 123–24). I also historicize both texts: I provide a brief history of the conquest of the Americas, with special emphasis on the large number of distinct tribal nations, the many differences among them, and the various forms Christianity took among Protestant and Catholic Europeans. To further prevent stereotyping, I make sure that the course syllabus includes a variety of Native and Judeo-Christian texts so that students cannot assume that *all* Native peoples or *all* Christians share the same views.

I also connect this reading unit with the course as a whole and explain that I have structured the syllabus as a series of nonlinear, interlocking transcultural dialogues among a variety of different texts and worldviews. I tell students that by requiring them to engage with multiple, interrelated perspectives I hope to encourage their relational thinking skills—their ability to employ analysis, imagination, and self-reflection in conjunction. Throughout the semester, we will use relational thinking as we investigate the ways literary texts, canons, and theories are shaped by personal, cultural, and historical issues. Drawing on Bakhtin's theory of language, I give students a rationale for my relational approach. I explain that language is never neutral; the words and texts we encounter are always context-specific and come to us already infused with meaning and value. By emphasizing the inevitably intersubjective nature of language, I hope to encourage my students to investigate how writers have appropriated and shaped the words they use to fulfill their own intentions—intentions that have themselves been influenced by external factors. As we move through the course, I explain, I want them to analyze and speculate on each writer's and each text's underlying assumptions, use their own knowledge and experience to analyze these worldviews, and draw connections among the various texts we read. I also inform students that, in my opinion, relational thinking assists them in achieving intellectual agency. As such, it can extend beyond the classroom and be useful in other areas of their lives: It can give them new perspectives about many issues, ranging from the advertisements they encounter every day to the stereotypes circulating in contemporary culture and media accounts of recent events.

My point here—and it's a point that I emphasize repeatedly to my students—is that I want them to be active learners. I briefly summarize the banking account model of education[8] and contrast it with my own relational, dialogic approach. I explain that if we simply accept what others tell us, we become victims: enslaved to others' ideas, inscribed by others' laws. For readers of this book, my brief summary of relational thinking and active learning comes as no surprise and indicates that my teaching has been influenced by scholarship in critical pedagogy. But for my undergraduate students, who have rarely (if ever) been challenged to reflect on their educational process, my emphasis on active learning is quite startling and gives them an opportunity to examine their unconsciously held assumptions about education. Moreover, by defining them as active learners, I inform them that I value their views and respect their intellectual ability. I find that this frank approach encourages students to be more open-minded and willing to engage themselves with the course material.

Once I have established the framework for our discussion, we are ready to explore similarities and differences between the Genesis chapters and "Talk Concerning the First Beginning." I request that students break into groups of four where they complete three tasks: (1) list the similarities and differences between the two narratives; (2) discuss what these similarities and differences might indicate about the texts' worldviews; and (3) speculate on how these worldviews might shape people's material circumstances (sociopolitical systems, gender relations, ethics, and so forth). As my use of the word *might* indicates, I underscore the tentative nature of their exploration; they should not assume that they will arrive at "the truth" about these issues. We then discuss their findings as a class. After generating lists of plot elements from each creation myth on the board, we explore what these elements might indicate about some Judeo-Christian and Native worldviews. We arrive at a number of insights. The Zuni narrative includes the following characteristics: an emphasis on communal identities; multiple non-anthropomorphic divine beings; many interactions among the divine, humans, and animals (who are depicted as relatives); and humans, animals, and the divine all play participatory roles in creation, which proceeds in cyclical fashion. These characteristics imply a holistic worldview positing nonhierarchical inter-relatedness and reciprocity among all dimensions of life. Applied to the material circumstances of people's lives, this holistic worldview could lead to egalitarian social arrangements, complementary gender relations, and political-economic systems valuing reciprocity, balance, and respect.[9]

The Genesis account is quite different. It contains a single anthropo-morphic, masculine divine being; distinct hierarchical divisions between the divine and the human, between human beings and animals, and between man (Adam) and woman (Eve); humans play a far more limited—and destructive—role in creation; the animals (unless we view that notorious serpent-defined-as-Satan as an animal) play no role at all; and creation moves from perfection into a state of exile and fallenness. These characteristics imply a hierarchical, dualistic worldview that despiritualizes nature and material reality. As Cajete explains,

> God the Creator was seen to live outside the universe, transcendent and greater than the universe, with dominion over the universe and all inhabitants. Humans had a connection to this divinity, but in order to fully consummate this union, they had to transcend the material world and exercise dominion over it in God's name. This orientation led to a perception of the world in purely material terms; hence, the objectification, secularization, and scientific orientation of the world. The nonhuman world (often including tribal Indigenous peoples) was the property of the transcendent God and his chosen people. Although it was holy, it was also material, without spirit, and therefore eligible for use or exploitation by God's chosen. This conception of the world as spiritless (dead/lifeless) material allowed Western peoples to have a sense of detachment that was religiously justifiable. Therefore, Western people could decide how they applied this God-given dominion over nature. (*Native Science* 280–81)

I would add that because "God the Creator" is depicted in distinctly masculine terms, the Genesis creation story can be used to valorize a patriarchal, phallocentric sociopolitical system. Moreover, the pairing of Adam and Eve, coupled with God's command that they be fruitful and multiply, gives the Genesis account what we could today describe as a divinely authorized heterosexual framework. Applied to the material circumstances of people's lives, this hierarchical worldview contributes to the development of rigidly structured social arrangements, unequal gender relations, and economic-political systems valuing authority and control. (See appendix 3.)

Generally by this point in our discussion, students will raise further objections. Disturbed by the negative implications of the Genesis text, they insist that we cannot fully comprehend its worldview without also including the New Testament story of Christ. I welcome this objection and invite students to discuss how expanding the Genesis creation account to include the Christ story alters their interpretations. Through this discussion, they recognize the important roles audience and context play in the interpretative process. I then ask the class to apply this objection to our interpretation of the Zuni story as well: Just as we must acknowledge that the Genesis account is part of a much larger story that the Puritans and other European colonizers bring to their interpretation of the Genesis myth, we must consider that the Zuni account is also part of a larger story, a story that *we do not know*. What are the implications of our ignorance? How might these knowledge gaps shape and alter our interpretations? These questions and the subsequent discussion challenge students to reflect more deeply and examine their own reading process. They can no longer approach "Talk Concerning the First Beginning" as a self-contained piece of literature; nor can they regard themselves as detached and therefore objective readers. Instead, they are compelled to acknowledge the biases and other limitations in their own perspectives and to make space for the unknown.

This exercise serves a number of purposes. First, as I've already explained, it introduces students to the relational, dialogic framework I use to structure the entire syllabus and all class discussions. This reading unit sets up a three-way potentially transformative dialogue among the worldviews of European colonizers, Zuni (Indigenous) inhabitants, and my students. As the semester progresses, the dialogue becomes increasingly complex. Thus for example we look at images of the Americas in texts by Columbus, Ann Bradstreet, Pedro Menéndez de Avilés, Gaspar Pérez de Villagrá, Handsome Lake, John Winthrop, Frederick Douglass, J. Hector St. John de Crèvecoeur, and Phillis Wheatley; we examine social protest in southwestern corridos and texts by David Walker, Harriet Jacobs, Henry David Thoreau, William Appess, Elias Boudinot, Hannah Webster Foster, and Susanna Haswell Rowson; and we investigate representations of selfhood and society in writings by Ralph Waldo Emerson, Samson Occom, Jonathan Edwards, Benjamin Franklin, Olaudah Equiano, and Margaret Fuller.[10] In each instance, we begin with a commonality and then move outward, exploring the ways history, economics, gender, ethnicity/race, and other variables influence the many different forms this commonality takes.

Second, and closely related, this exercise lays the thematic groundwork for the entire semester, enabling students to better understand a variety of literary and historical issues, including (but not limited to): the colonizers' view of nature as a "howling wilderness" that must be subdued; the Puritans' paradoxical sense of themselves as simultaneously wretched, fallen, utterly worthless creatures yet (because made in God's image) entitled to and divinely authorized to dominate nature—along with the Indigenous peoples they so conveniently conflated with the natural world; the focus on self-contained individualism central both to U.S. culture and to the conventional literary canon;[11] the emphasis on community and self-in-relation often found in non-canonical writings; and the Transcendentalists' attempts to develop holistic philosophical systems. Third, this exercise challenges students' stereotyped views of "Indians" as primitive savages by demonstrating that Native peoples have distinct, highly sophisticated cosmological systems that (certainly in this exercise) seem far more humane (and "civilized") than the worldviews embedded in the Genesis creation account.[12] Through class discussion, I highlight this challenge to stereotypes. After tracing the many ways Indigenous relational worldviews have been distorted into romanticized stereotypes about "Indians" and their automatic proximity to "nature," I explain how "New Age" consumer culture continues this appropriation.[13]

Perhaps most importantly, given my goals as an educator, this exercise enables me to expose the unspoken mainstream norm, thus encouraging students to reflect on their own worldviews and introducing them to alternative perspectives. This self-reflection is a crucial component of transformational multiculturalism. Many of my students have been so immersed in conservative interpretations of Judeo-Christian beliefs that they accept this hierarchical, dualistic status-quo story as an absolute truth— the only possible description of reality. Like Cajete, I believe that this Judeo-Christian cosmology, as generally interpreted,[14] is "dysfunctional . . . [and] can no longer sustain us at any level" (*Native Science* 281). Through class discussion, I tease out the "dysfunctional" dimensions: this hierarchical worldview separates nature (defined broadly to include animals, vegetation, and the land itself) from human life, subordinating the former to the latter. It naturalizes and divinely authorizes a system that ranks human beings according to ancestry, economic status, and gender. It dehumanizes entire populations of people and justifies their enslavement, destruction, and other horrific abuses. It normalizes an unmarked 'white' norm; and it fosters a hyper-individualism that values competition over reciprocity.

However, it is not enough just to expose the dysfunctional elements. As Audrey Thompson notes, "we cannot simply dismantle students' worlds, while putting nothing new in their place. We cannot give them tools for critiquing and deconstructing oppression without giving them something else to strive toward" (445). By introducing students to relational worldviews, connectionist thinking, and other concepts related to holism, I offer them alternative perspectives from which they can, if they so choose, view themselves and their worlds. Of course, I cannot *force* my students to change their views. Nor would I wish to do so. My goals are more modest: I want to expose, denaturalize, and historicize both the dominant-cultural framework and the status-quo stories[15] it holds in place. I challenge students to explore the ways this framework shapes present-day U.S. culture and influences their own worldviews as well as their reading practices. By so doing, I heighten their sense of agency and hold them more accountable for the choices they make.

Relational teaching is situation-specific and can take many forms. These forms, in turn, will be shaped by the courses we teach; by our own identities, interests, experiences, and concerns and those of our students; and by other variables. I don't know if my "creation myths" exercise would work with students who have not been immersed in fundamentalist Christian beliefs, and I doubt it would be appropriate in a course on, say, twentieth-century British literature. However, no matter what specific form relational teaching takes, in each instance we begin with commonalities, which themselves are always context-specific. These commonalities neither overlook nor deny the differences among us. Rather, they offer pathways into relational investigations of difference—difference defined not as deviation *from* an unmarked norm but as interrelated *with* this norm. This nonbinary approach to difference, coupled with (sometimes painful) self-reflection, makes transformation possible, on both individual and collective levels.

As a number of theorists suggest, the narratives we read—the stories we tell ourselves and our students—play an important role in constructing "American" identities and themes.[16] Perhaps not surprisingly, the highly celebrated belief in a fully independent "American" self has, to a great degree, shaped the ways we see ourselves and our world. In this story, the boundaries between self and other become rigid, inflexible, and far too restrictive. Significantly, these self-enclosed, isolated identities are themselves produced by an oppositional model of identity formation authorized in dominant-cultural philosophical and psychoanalytic accounts of subjectivity. Yet as Kelly Oliver asserts, this oppositional model normalizes "war, hatred, and oppression," making them "inevitable and unavoidable parts of social development" ("Identity" 175). The relational worldviews presented by Indigenous teachings provide a different narrative, one with important consequences for contemporary social life. As Cajete explains,

> What traditional Native models have to offer is a perspective of community that goes beyond the social "isms" and theories of community. Native models of community get to the heart of social relationship as an expression of human biophilia, and of human society as a part of nature rather than separate from nature. This ancient idea of relationship must be allowed to arise in our collective consciousness once again. In this perilous world of the twenty-first century, it may well be a matter of our collective survival. (*Native Science* 105)

Relational models of community and relational worldviews occur in non-Native texts and cultures as well. However, given the pedagogical objectives I've described in this chapter, I have found it extremely effective to use Native American literature as entry points into relational perspectives.

permeable boundaries

As I stated in chapter one, language, belief, perception, and action are intimately interrelated. How we perceive ourselves, other people, and the world affects how we act. My experiences have taught me that the recognition of our profound interconnectedness offers a vital key to long-term individual/collective change, a crucial point of departure in my work for social justice: if we are all interconnected, then the events and belief systems impacting my sisters and brothers in New York City or Baghdad or Jerusalem have a concrete effect on me. Like Anzaldúa, "I have come to realize that we are not alone in our struggles nor separate nor autonomous but that we—white black straight queer female

male—are connected and interdependent. We are accountable for what is happening down the street, south of the border or across the sea" ("Foreword" n.p.).

Immersed in status-quo stories of possessive individualism, the majority of my students cannot recognize their interconnections with or accountability for others. The relational teaching approach I have described in this chapter and will elaborate on in the following pages enables me to expose and challenge my students' fragmented, solipsistic beliefs and the underlying narrative buttressing these beliefs. When we read, we enter into the worldviews of others. Though these journeys are always partial, always incomplete, they can affect us profoundly. As they travel into the worlds (the lives, the perspectives, the beliefs, the histories . . .) of their various 'others,' students must (re)examine their own worldviews. Perhaps their protective boundaries between self and other will become permeable, begin breaking down. Perhaps those of my students who become teachers will not offer their own students simplistic narratives of historical events such as the commemorative "First Thanksgiving Day" my daughter learned at school. Perhaps they will design situation-specific, relational teaching tactics that acknowledge the ways we are mutually implicated in historical and contemporary events. This, at least, is my hope and one of my goals as an educator.

CHAPTER THREE

Giving Voice to 'Whiteness'? (De)Constructing 'Race'

Sticks and stones may break our bones, but words—words that evoke structures of oppression, exploitation, and brute physical threat—can break souls.

Kwame Anthony Appiah[1]

"Racing" is a practice of separating people out from the general population with the specific purpose of fortifying the dominance of the remaining majority. Thus, race is not a passive recognition of natural qualities, but rather the sum of intentional actions taken to stratify the population in order to maintain white privilege and non-white subordination.

john a. powell[2]

Regardless of what I might experience professionally or personally, I believe that race consciousness hinders, if not destroys, us all. We cannot liberate ourselves by using race.

Reginald Leamon Robinson[3]

I am deeply suspicious about the ways 'race' functions in contemporary U.S. culture. As this chapter's epigraphs suggest, I believe that our racialized thinking and speaking, although perhaps unavoidable, lock us into destructive, soul-breaking patterns. 'Race' categories are built on a series of brutal, exclusionary practices originating in histories of oppression, manipulation, land theft, body theft, soul theft, physical and psychic murder, and other crimes against specific groups of people. These categories were motivated by economics and politics, by insecurity and greed—not by innate biological or divinely created differences. 'Race' has a poisonous history that continues infecting us today. Every time we automatically refer to 'race' or to specific 'races' we draw on and thus reinforce this violent history, as well as the 'white' supremacism buttressing the entire system.

To be sure, positive racial discourse has been vital, enabling people of colors to gain self-affirming historical and sociopolitical agency and in other ways make sense of our lives. I know from personal experience how important and helpful racial categories can be in the face of alienation, discrimination, ostracism, or other forms of isolation and bias. As Houston Baker, Jr., explains, 'race' offers "a recently emergent, unifying, and forceful sign of difference *in the service* of the 'Other.' " He explains that for people of colors racial identities function as "an inverse discourse— talk designed to take a bad joke of 'race' . . . and turn it into a unifying discourse" (386, his emphasis). Although Baker acknowledges the destructive, fictionalized aspects of 'race' (it is, after all, a "bad joke"), he maintains that African Americans and other racialized groups can reverse its negative implications and use racial discourse in affirmative ways, ranging from enhanced self-image and increased self-esteem to social transformation.

However, this oppositional use of 'race' is too limited, for it cannot successfully transform the negative energies, motivations, and assumptions underlying *all* references to 'race.' Even highly affirmative talk of a black, or Chicano/a, or Native American racial identity emerges from and reinforces already-existing, destructive ideologies of 'race.' john a. powell makes a similar point: "The celebration of an identity that is significantly defined and shaped by racist oppression without adopting, to some extent, the terms of the oppressor is a difficult, if not impossible, enterprise" ("Our Private Obsession" 120).[4] Our references to 'race' participate in and empower status-quo stories that have functioned historically to create dehumanizing, hierarchical divisions based on false generalizations concerning physical appearance and other arbitrary characteristics.[5]

By thus reinforcing monolithic stereotypes and other inaccuracies, contemporary racialized discourse creates further divisions among people. As Henry Louis Gates, Jr., points out,

> The sense of difference defined in popular usages of the term "race" has both described and *inscribed* differences of language, belief system, artistic tradition, and gene pool, as well as all sorts of supposedly natural attributes such as rhythm, athletic ability, cerebration, usury, fidelity, and so forth. The relation between "racial character"

and these sorts of characteristics has been inscribed through tropes of race, lending the sanction of God, biology, or the natural order to even presumably biased descriptions of cultural tendencies and differences. ("Writing 'Race' " 5, his emphasis)

Despite the many historic and contemporary alterations in racial categories, people generally treat 'race' as an unchanging, biological fact. Often, they make simplistic judgments and gross overgeneralizations based primarily on outer appearance. You know the stereotypes: "All whites are bigots." "Blacks are more athletic than other people, and boy can they dance!" "All 'Hispanics' are hot-blooded." "All Asians excel at science and math." In addition to perpetuating untruths, these stereotyped, racialized beliefs imply a homogeneity within each so-called 'race.'

I have not always viewed 'race' with such deep hostility and suspicion. In fact, for years I believed (and lived according to) the status-quo stories about 'race.' But my experiences in the classroom for the past twelve years have transformed my perception of 'race.' In the following pages, I retrace part of my journey.

defining 'whiteness'?

During my first few years as a teacher, I carefully incorporated issues related to gender, economic status, 'race,' sexuality, and other categories of human difference into my syllabi and class discussions. Drawing on recent scholarship, I encouraged students to examine the various ways these categories impact U.S. culture and literary texts. When I taught Leslie Marmon Silko, N. Scott Momaday, or Paula Gunn Allen, for example, I described the authors' perspectives on contemporary Native American literary and cultural conventions and asked students to consider the various ways their poetry, fiction, and prose seemed to reflect and reshape these standards. After an initial period of questioning, students generally arrived at useful and interesting observations. Similarly, when I taught Nella Larsen and Paul Laurence Dunbar, I described the status of "Negroes" in the early 1900s and asked students to consider how Larsen's and Dunbar's 'race' might have influenced their work. Again, students arrived at insightful comments about the texts. Although this approach generally worked quite well, I had a nagging sense of imbalance . . . something was lacking. But what? After reading Toni Morrison's *Playing in the Dark: Whiteness and the Literary Imagination*, I realized that discussions of 'whiteness'—*as* 'whiteness'—had been missing from my course.

In her groundbreaking book, Morrison calls for an examination of 'whiteness' in canonical U.S. American literature. What, she asks, are the implications of "literary whiteness" (xii)? How does it function in the construction of an "American" identity? Arguing that "[a] criticism that needs to insist that literature is not only 'universal' but also 'race-free' risks lobotomizing that literature, and diminishes both the art and the artist" (12), she urges scholars to examine the hidden racial discourse in U.S. literature.

I was delighted by Morrison's call for an interrogation of literary 'whiteness.' Of course! It only made sense to include analyses of 'whiteness' in class discussions of racialized meanings in literary texts. After all, we were already examining black, Latino/a, Native American, and Asian American literary traditions. By not also investigating 'white' traditions *as 'white,'* we were converting literary 'whiteness' into the norm.[6] In this schema, "minority" writings become deviations from the unmarked ('white') norm. This unbalanced labeling process distorts students' perceptions, creating a hierarchal relationship among authors—with 'white'-raced authors at the top. Gloria Anzaldúa makes a similar point while discussing the implications of the fact that she is rarely referred to simply as "author" but instead has been labeled a "*Chicana* author": "Adjectives are a way of constraining and controlling. . . .The adjective before writer marks, for us, the 'inferior' writer, that is, the writer who doesn't write like them. Marking is always 'marking down' " ("To(o) Queer" 250). This marking process also reinforces students' already-existing, although often unconscious, belief that 'white'-raced authors are "just human" whereas all others, because racially marked, are inferior or in some other way abnormal.

And so, shortly after reading *Playing in the Dark*, I decided to incorporate explorations of 'whiteness' into the U.S. American literature courses I teach, starting with my fall 1993 survey of nineteenth- and twentieth-century literature. The majority of the students taking this course were working-class, first-generation college students in their early twenties. Over two-thirds self-identified as "white;" and the remaining students self-identified as "Hispanic," "Spanish," "Pueblo," "Navajo," and "black." These students' reactions greatly shaped the next steps on my journey through 'race.'

teaching 'whiteness'?

I was exhilarated when I began including investigations of 'whiteness' in the survey course, and I was confident that our examination would go well. I had already laid the groundwork by discussing the various ways language is racialized and suggesting that literature can be examined for its hidden and overt racial meanings. However, when I applied this approach to literary 'whiteness,' students (of all colors) were stunned into silence. I began by suggesting that 'whiteness' can be viewed as a *racialized* identity, continually reinforced and reinvented in literature. Students met this assertion with disbelief: People with pale skin are often referred to as "whites," and of course there are ethnic groups whose members have so-called 'white' skin—Italian Americans, Polish Americans, many U.S. Jews, and so on—but a white 'race'?[7] Although I discussed Morrison's call for an interrogation of literary 'whiteness' at great length, when I asked students to speculate on how Herman Melville, John Greenleaf Whittier, and other 'white'-raced authors have contributed to this tradition of literary 'whiteness,' they were puzzled and unable to reply. Nor could they discuss Hawthorne's 'whiteness,' or analyze how Caroline Kirkland's 'race' might have influenced *A New Home—Who'll Follow?* Clearly, my students had no idea about what this 'whiteness' might entail. But to be honest, my own ideas about 'whiteness' were becoming slippery and difficult to pin down.

Classroom investigations of 'whiteness' were especially tricky and confusing when we focused exclusively on 'white'-authored texts that contained almost no explicit references to 'race.' How should we analyze 'whiteness' in Emerson's essay on "Self-Reliance"? Should we code key themes in "Self-Reliance"—themes such as the desire for personal agency, a sense of inspired self-confidence, feelings of spiritual connection with nature and the divine, belief in the importance of creating one's own community—as 'white'? I felt uncomfortable identifying these positive qualities as attributes of literary 'whiteness,' for to do so leads to significant problems when we encounter these 'white' themes in texts by authors of color. If, for example, the quest for independence and self-trust is coded as 'white,' must we suggest that in his *Narrative* Frederick Douglass becomes or acts 'white' when he asserts his intellectual independence from Covey; or when, after his failed attempt at a communal escape, he resolves to "trust no one"? Such rigid assumptions cannot facilitate increased understanding of the literature we read or the world in which we now live.

And so, I modified my focus. Rather than examining *literary* 'whiteness' as illustrated by 'white'-raced authors, I told the class, we would examine representations of 'whiteness' in texts by authors of colors. Although this approach untied my students' tongues, it was not very effective. Despite my repeated reminders that we must distinguish between literary representations of 'whiteness' and real-life people classified as 'white,' students of all colors found it extremely difficult (and at times impossible) not to blur the boundaries between them. Although I generally champion the blurring of boundaries, in this pedagogical instance blurred boundaries were quite detrimental. Students became immersed in highly negative, overly generalized explorations of 'white' people. Class discussion of "School Days of an Indian Girl," an autobiographical narrative by Zitkala-Sä, an early twentieth-century mixed-blood writer, illustrates this transition from 'whiteness' to 'white' people. Although they could analyze the ways Zitkala-Sä depicted her entrance into the 'white' world, students seemed reluctant to take this analysis further by speculating on what these depictions might tell us about representations of literary 'whiteness.' Instead, they focused their attention on Zitkala-Sä's representations of 'white' people, which they conflated with all real-life human beings classified as 'white.' Their analyses portrayed 'white'-raced people in an entirely negative light: 'whites' were emotionally and spiritually cold; overly concerned with rules and order; rude and disrespectful; and prideful and extremely racist, especially in their negation of Native cultures, peoples, and beliefs.

Given the historical context of Zitkala-Sä's narrative—the U.S. government's attempts to forcibly remove, exterminate, assimilate, reeducate, sterilize, and/or Christianize Native peoples—my students' demonization of 'whiteness' is not surprising. Yet despite my repeated reminders, they made almost no distinction between literary 'whiteness' and all 'white'-raced people. Instead, they created simplistic binary oppositions between the "good Indians" and the "bad whites."[8] Like all binary oppositions, this dualism is far too simplistic and conflates literary representations of 'whiteness' and 'white' people with a homogeneous, ahistorical group composed of all real-life human beings racialized and classified as 'white.' As I will explain in the following chapter, this conflation can have unexpected, unproductive effects on some students who identify as 'white.' These students experience a variety of reactions, ranging from a self-centered, paralyzing guilt to 'white'-supremacist racialization as the newest oppressed "minority."

These early classroom investigations of 'whiteness' pointed to other closely related problems as well. How do we separate 'whiteness' from masculinity and other forms of privilege and entitlement, or is it even possible to do so? Perhaps Owen J. Dwyer and John Paul Jones III are correct when they argue that we cannot fully distinguish 'whiteness' from other forms of privilege, insisting that "whiteness is not distinct from either colonialism and masculinity" (210). Is it 'whiteness,' masculinity, 'white' masculinity, or some other combination of these elements and perhaps others as well that enables Emerson, Thoreau, *and* Douglass to attain remarkable levels of confidence and self-assertiveness in their prose? In our discussions of individualism in Emerson's and Thoreau's writings, some students assumed that both writers came from wealthy backgrounds and insisted that it must be class privilege, rather than 'whiteness,' which enabled these men to achieve self-reliance. Given the financial hardships both writers experienced at various points in their lives, this suggestion, though plausible, oversimplifies.

The further we traveled in our investigation of 'whiteness,' the greater my questions grew. What is the relationship between those people racialized as 'white' and literary 'whiteness'? For instance, should we assume that because Walt Whitman was of European descent and therefore 'white,' his writings give us insight into literary 'whiteness'? Should we place his writings in a canon of 'white' U.S. literature? After all, this practice of categorizing literature according to the author's 'race' has played a pivotal role in scholars' attempts to construct African American, Native American, and other cultural/ethnic-specific canons. But if we categorize literary 'whiteness' in this way, 'whiteness' becomes a synonym for 'white'-raced people—making "literary 'whiteness' " just a fancy way of saying "anything written by 'whites' "—which is certainly not what I meant when I encouraged my students to investigate literary 'whiteness'! To be sure, equating literary 'whiteness' with 'white' people exposes the 'white'-supremacist status quo; however, exposure and transformation are not synonymous. Should I, then, define 'whiteness' more broadly—as a culture, perhaps? If so, who belongs to and inhabits this 'white' culture? Who shapes its values, ethics, and beliefs: *All* people racialized as 'white,' or only some of these people? (And if so, which ones?) Where do so-called mixed-'race' people and assimilated people of colors fall in relationship to these various forms of 'whiteness'?

These questions and my unsuccessful early attempts to incorporate 'whiteness' into classroom discussions propelled me into further research on 'whiteness,' 'whiteness studies,' and 'white' people. As I explain in the following section, I discovered that scholars generally read 'whiteness' in dialogue with 'blackness' and describe the former in almost entirely negative terms.

marking 'whiteness'

My students and I were not alone in our inability to fully comprehend, name, and explain 'whiteness;' as Kobena Mercer states, "[o]ne of the signs of the times is that we really don't know what 'white' is." Thus he asserts that "the real challenge in the new cultural politics of difference is to make 'whiteness' visible for the first time, as a culturally constructed ethnic identity historically contingent upon the disavowal and violent denial of difference" (205–06). Since the early 1990s a growing number of scholars in cultural studies, history, literary studies, geography, women's studies, and other disciplines have seized this challenge. Building on the work of ethnic studies and critical 'race' theory, scholarship in 'whiteness' studies takes a variety of approaches, ranging from investigations of historical and contemporary forms of regional-, class-, and gender-specific 'white' identity to analyses of 'whiteness' in film, literature, and other cultural texts.[9]

To be sure, when not conducted with great care 'whiteness' studies scholarship becomes a type of self-centered navel-gazing: yet another way to keep the focus on 'white'-raced people. As Margaret L. Andersen explains, "In the absence of a material analysis of race and racialization, much of the whiteness studies scholarship devolves into white identity politics" (38). Rather than contribute to the dismantling of racism, these narrow, solipsistic versions of 'whiteness' studies detract from anti-racist work. Michael Apple makes a similar point, warning that the focus on 'whiteness' "run[s] the risk of lapsing into the possessive individualism that is so powerful in this society . . . [S]uch a process can serve the chilling function of simply saying 'But enough about you, let me tell you about me'. Unless we are very careful and reflexive, it can still wind up privileging the White, middle class woman's or man's need for self-display" ("Freire" 115).

Despite these risks and others,[10] investigations of 'whiteness' should not be dismissed as one more scholarly fad in academia's "publish-or-perish" game. 'Whiteness' has functioned—in education, the media, U.S. politics, and many other areas of life—as a pseudo-universal category that hides its specific values, epistemology, and other attributes under the guise of a nonracialized, supposedly colorless, "human nature." Its particular presence erased, this naturalized, unspoken 'whiteness' (mis)shapes contemporary western cultures and masks social and economic inequalities. By negating those people—*whatever* the color of their skin—who do not measure up to 'white' standards, 'whiteness' has played a central role in maintaining and naturalizing a hierarchical, racist social system and a dominant/subordinate worldview. As such, it is vital that we expose and interrogate 'whiteness' while avoiding the possessive individualism Andersen and Apple describe.

Although academic 'whiteness' studies has grown tremendously in the past two decades, the concept of 'whiteness' remains somewhat elusive.

Whether scholars try to define or describe it, 'whiteness' elides definition yet has multiple, overlapping, sometimes-conflicting meanings. Paul C. Taylor makes a similar point in his discussion of 'whiteness studies,' noting that "the 'core concept—whiteness—fairly defies singular definition.' " Although he has read extensively in the field, he has found that "the literature . . . has been more suggestive than clear. Where one might expect to find definitions, one often finds a metaphor or list of similes: whiteness-as-property, whiteness-as-terror, whiteness-as-invisibility, and so on" (227, 228).[11] Scholars try to mark 'whiteness,' but they cannot pin it down and assign it fixed meanings; 'whiteness' slips through their gaze, slides through their words.

Not surprisingly, then, one of the most commonly mentioned attributes of 'whiteness' is its pervasive nonpresence, its invisibility. A number of scholars associate this ubiquitous hidden 'whiteness' with unmarked forms of superiority and power. As Richard Dyer suggests in "White," his influential early analysis of representations of 'whiteness' in mainstream U.S. and British film, "white power secures its dominance by seeming not to be anything in particular" (44). Drawing on scientific studies of chromatics, he explains that whereas black—because it is always marked as a color—refers to particular objects and qualities, white does not: It "is not anything really, not an identity, not a particularising quality, because it is everything—white is no colour because it is all colours" (142). In literary and cultural studies this "colourless multicolouredness" gives 'whiteness' a pseudo-universality and omnipresence quite difficult to examine: "It is the way that black people are marked as black (are not just 'people') in representation that has made it relatively easy to analyse their representation, whereas white people—not there as a category and everywhere everything as a fact—are difficult, if not impossible, to analyse *qua* white" (143, his emphasis).

This association between 'whiteness' and invisibility can apply to 'white'-raced people as well. According to Crispin Sartwell, who self-identifies as 'white,' "white race consciousness" is entirely lacking in self-reflexivity. Many people classified as 'white' have absolutely no idea of what it means to be 'white':

> it would be not too much to say that white race consciousness *is* race unconsciousness, or that our white race consciousness is not a consciousness of ourselves, but precisely of *you*. I've actually tried asking white people *what it means to be white*, only to meet with blank stares. It means nothing, phenomenologically, to be white; to be white is to be whited out; white consciousness is the erasure of

itself. White people will say there is no content to whiteness, even as they are able to produce amazingly detailed knowledge of what is not-white, that is, what they must avoid to stay white. This erasure is precisely the phenomenological form of white authority, its formation of itself as the position of no position: the position of objectivity to which epistemic and economic mastery accrues. (140–41, his emphasis)

I have seen similar dynamics in the classroom when students who are classified as 'white' insist that being 'white' has absolutely no meaning for them. This extreme lack of self-reflection, coupled with "the phenomenological form of white authority," significantly shapes the destructive forms of individualism I discussed in chapter one. As Martha R. Mahoney explains, "Because the dominant norms of whiteness are not visible to them, whites are free to see themselves as 'individuals,' rather than as members of a culture" ("Social Construction of Whiteness" 331). To be sure, there is nothing wrong with viewing oneself as an individual! After all, each of us is a distinct human being. The problem arises because they do not see non-'white' people as individuals but instead view them as racially marked and therefore members of specific racialized groups.

The invisible omnipresence of 'whiteness' has given it (and many 'white'-raced people) a too-rarely acknowledged position of dominance and power with material socio-cultural effects. As Henry Giroux suggests, 'whiteness,' domination, and invisibility are almost synonymous. He asserts that although " 'whiteness' functions as a historical and social construction," the dominant culture's inability or reluctance to see it as such is the source of its hidden authority; 'whiteness' is an unrecognized and unacknowledged racial category "that secures its power by refusing to identify" itself ("Post-Colonial Ruptures" 15).[12]

Thus many scholars associate 'whiteness' with superiority and control. Frye identifies what she calls "whiteliness" ('white' ways of thinking and acting) with the desire for personal and collective power, asserting that "[a]uthority seems to be central to whiteliness, as you might expect from a people who are raised to run things" ("White Woman Feminist" 156). She describes "whitely" people as "judges" and "preachers" who—because they assume that their "ethics of forms, procedures and due processes" represent the *only* correct standard of conduct—attempt to impose their beliefs on all others (155). Dyer makes a related point in his discussion of *Simba*, a 1955 colonial adventure film depicting the conflict between British colonizers and the Mau Mau in Kenya, where 'white' is coded as orderliness, rationality, and control, in contrast to 'black,' which is coded as chaos, irrational violence, and loss of control. Morrison notes a similar pattern of restrictive 'white' thinking, which she associates with an insistence on purity, self-containment, and impenetrable borders. According to Morrison, 'white' literary representations establish "fixed and major difference where difference does not exist or is minimal." For instance, metaphoric references to "the purity of blood" have enabled writers to construct a rigid, inflexible division between 'white' civilization and 'black' savagery (68). This division plays itself out in numerous works of U.S. literature, where differences based on blood[13] are used to empower 'white' characters.

With its presence erased, 'whiteness' functions as the unacknowledged standard against which all so-called "minorities" are measured, found lacking, and marked as deviations from this (unmarked) 'white' norm. As Dyer explains, "[l]ooking, with such passion and single-mindedness, at non-dominant groups has had the effect of reproducing the sense of oddness, differentness, exceptionality of these groups, the feeling that they are departures from the norm. Meanwhile the norm has carried on as if it is the natural, inevitable, ordinary way of being human" ("White" 44).[14]

This invisible, omnipresent, naturalized 'white' norm has led to a highly paradoxical situation: On the one hand, it is vital that we examine the many ways 'whiteness' has shaped U.S. culture; on the other hand, its pervasive nonpresence has made it difficult to analyze 'whiteness' as 'whiteness.' As Dyer asserts, "if the invisibility of whiteness colonises the definition of other norms—class, gender, heterosexuality, nationality and so on—it also masks whiteness as itself a category" ("White" 46). Consequently, theorists of all colors have adopted a relational approach, where 'whiteness' is examined in the context of 'blackness' or (more rarely) other racialized categories. Like some of my own early classroom attempts to explore literary 'whiteness,' this approach does not focus exclusively on 'whiteness' but instead examines 'whiteness' as it appears in the context of racialized others. In "White Woman Feminist," for example, Marilyn Frye draws on U.S. women of colors' and African Americans' discussions of 'white' people to explore "whiteliness."[15] Dyer centers his analysis of 'whiteness' in mainstream cinema around instances where the narratives "are marked by the fact of ethnic difference" ("White" 46).

Morrison takes a similar approach in *Playing in the Dark*, where she maintains that literary 'blackness'—or what she terms "Africanisms"—are central to any investigation of literary 'whiteness.' She begins with the hypothesis that "it may be possible to discover, through a close look at literary 'blackness,' the nature—even the cause—of literary 'whiteness'" (9).

Throughout her book she explores literary 'whiteness' by examining how "notions of racial hierarchy, racial exclusion, and racial vulnerability" influenced 'white' writers "who held, resisted, explored, or altered those notions" (11). For instance, in her discussion of Willa Cather's *Sapphira and the Slave Girl*—which depicts the interactions between Sapphira, a 'white' slaveholder, and Nancy, an enslaved woman Sapphira "owns"— Morrison examines the ways 'white' womanhood acquires its specific identity, as well as its power, privilege, and prestige, at the expense of 'black' womanhood. And in her examination of *Huckleberry Finn* she demonstrates that the notions of independence and freedom in this novel rely on the presence of the unfree Jim.

Like Morrison, Aldon Lynn Nielsen focuses his analysis of literary 'whiteness' on the (racist) ways 'white' writers depict 'blackness.' In *Reading Race: White American Poets and the Racial Discourse in the Twentieth Century*, he associates 'whiteness' with a racist symbolic system deeply embedded in U.S. thinking and explores how 'white' identity has been constructed through racist stereotyping of the 'black' other. More specifically, he examines what he terms "frozen metaphors," or stereotypes of "blacks" that reinforce "an essentially racist mode of thought," privileging people of European descent while relegating people of African descent to an inferior position (3). In the numerous racist stereotypes he describes, representations of 'blackness' take a variety of sometimes contradictory forms yet have one thing in common: in each instance, they exist to affirm the validity and power of 'whiteness' (and, by extension, 'white'-raced people). By depicting people of African descent as lazy, carefree, unsophisticated, and primitive, Hart Crane, e.e. cummings, T.S. Eliot, and many other twentieth-century 'white' writers locate 'blackness' outside western cultural traditions, thus reinforcing already-existing beliefs concerning the superiority of 'white' aesthetics.

Perhaps not surprisingly, given the difficulty in focusing specifically on 'whiteness,' theorists also associate it with mystery, absence, and death. Morrison explains that although representations of 'blackness' serve a variety of symbolic functions in U.S. American literature, "[w]hiteness, alone, is mute, meaningless, unfathomable, pointless, frozen, veiled, curtained, dreaded, senseless, implacable" (*Playing* 59). Dyer, in his exploration of mainstream cinema, finds that on the infrequent occasions "when whiteness *qua* whiteness does come into focus, it is often revealed as emptiness, absence, denial or even a kind of death" ("White" 44, his italics). In *Night of the Living Dead*, for instance, all 'white' people are closely associated with death: "Living and dead whites are indistinguishable, and the zombies' sole raison d'être, to attack and eat the living, has resonances with the behaviour of the living whites" (57). According to hooks, these literary and filmic representations of 'whiteness' as mystery and death reflect a common belief in African American communities; during her own upbringing, she explains, "black folks associated whiteness with the terrible, the terrifying, the terrorizing. White people were regarded as terrorists" (*Black Looks* 170).

In this passage, hooks moves seamlessly from 'whiteness' to 'white' people, making the two synonymous. Like my students during our discussion of Zitkala-Sä described earlier in this chapter, hooks seems to be carried away by the proximity between 'whiteness' and 'white'-raced people. This conflation of the two, though often appropriate and analytically useful, can be dangerous. 'Whiteness' has so many negative characteristics that it is difficult (if not impossible!) to view it in a positive light. But what are the consequences of automatically applying this morally bankrupt, wretched 'whiteness' to all people racialized as 'white'?

Like other stereotypes, this conflation of 'whiteness' with 'white'-raced people draws on false generalizations and implies that all human beings classified as 'white' *automatically* exhibit the many negative traits associated with 'whiteness': They are, *by nature*, insidious, superior, empty, terrible, terrifying, dominating, controlling, superficial, empty, carriers of death, and so on. Now, I know 'white' folk who aren't like this; and whereas I would definitely agree that 'white' skin and at least some of these 'white' traits are often found together, I believe that the relation between them is conditional. As Frye suggests, "the connection between [whiteness] and light-colored skin is a *contingent* connection: this character could be manifested by persons who are *not* white; it can be absent in persons who are" ("White Woman Feminist" 151–52, her emphasis). In other words, the fact that a person is born into a family of European ancestry and thus racialized as 'white' does not necessarily mean that s/he will think, act, and write in the 'white' ways I've described. Nor does the fact that a person was not born into a family of European ancestry automatically guarantee that s/he will *not* think, act, and write in 'white' ways.[16] Leslie Marmon Silko beautifully illustrates this contingent nature of 'whiteness' and skin color in *Ceremony*, where full-blood Native characters such as Emo, Harley, and Rocky think and act in 'white' ways. As I will explain in more detail in the following chapter, although she too demonizes 'whiteness'—associating it with greed, restrictive boundaries, destruction, emptiness, absence, and death—Silko does not automatically associate 'whiteness' with all 'white' people. Indeed, it is the light-skinned, part-'white,' mixed-blood protagonist, Tayo, who learns to recognize and resist this evil 'whiteness.'

Unfortunately, however, as I have learned from my own experiences researching and teaching 'whiteness,' it becomes extremely difficult *not* to equate the word 'whiteness'—and, by extension, the many negative qualities it implies—with all 'white'-raced people. In fact, as I became deeply immersed in my research on the scholarship of 'whiteness' studies, I found myself making automatic assumptions about everyone who looked 'white.' I became uncomfortable and distrustful around those students and acquaintances I classified as 'white;' during this early stage in my own interrogation of 'whiteness' I was tempted to draw on my African ancestry, disavow my 'white' education, and entirely separate myself (intellectually, if not physically) from the so-called 'white race.' But of course this type of disavowal is impossible, unhealthy, and extremely unwise.

My exploration of the scholarship on 'whiteness' did not improve classroom investigations. Although I still believed that it was crucial to examine 'whiteness' whenever examining any racialized identities, I remained puzzled about how to do so without inadvertently reinforcing stereotypes, boundaries, and other arbitrary divisions. I still did not know how to incorporate 'whiteness' into my teaching in ways that could challenge and potentially transform students' perceptions. For my students and I, the relational approach employed by Dyer, Morrison, Frye, Nielsen, and many others was ineffective because it led to a series of binary oppositions between 'of color' and 'white.' And so, I again shifted my focus. Rather than exclusively researching 'whiteness,' I began investigating 'race' more generally. As I explain in the following section, delving deeply into the history of 'race' radically altered my perspective. It stripped away my assumptions but left me with additional questions and concerns.

(de)constructing 'race'

Like 'whiteness,' 'race' is an ambiguous, slippery concept that cannot be entirely contained in or circumscribed by scientific descriptions. As Michael Omi and Howard Winant persuasively demonstrate, "[t]he meaning of race is defined and contested throughout society, in both collective action and personal practice. In the process, racial categories themselves are formed, transformed, destroyed, and re-formed" (61). Even a brief look at a few of the many ways racial groups have been redefined in this country illustrates how *unstable* and *artificial* racialized identities are.

Throughout the nineteenth century many U.S. state and federal agencies recognized only three 'races,' which they labeled "White," "Negro," and "Indian." Given the extremely diverse mixture of people living in the United States, this three-part classification was, to say the least, confusing. How were people of Mexican or Chinese ancestry to be described—as "white"? "Negro"? or "Indian"? The state of California handled this dilemma in a curious way: Rather than expand the number of 'races,' the government retained the existing categories and classified Mexican Americans as a 'white' population and Chinese Americans as "Indian." As Omi and Winant explain, this decision had little to do with outward appearance; it was motivated by socioeconomic and political concerns, for it allowed the state to deny the latter group the rights accorded to people classified as 'white' (76).

Since then, both groups have been redefined numerous times. At various points in the last hundred years, U.S. Americans of Chinese descent have been classified as "Orientals," "Asians," "Asian Americans," "Pan Asians," and "Asian Pacific Americans." Yet each of these terms is inadequate and erroneously implies a homogeneity unwarranted by the many nationalities, geographical origins, languages, dialects, and cultural traditions forced into these politically motivated categories. As Yehudi Webster notes, these monolithic labels indicate the U.S. government's attempt to group "heterogeneous populations into one category on the basis of apparent similarities in skin color, hair type, and eye shape" (132–33).

Efforts to classify U.S. Americans of Mexican ancestry have been at least equally unsuccessful. Before 1930, the U.S. Census Bureau classified them as "white." In 1930, during a time of increased national xenophobia, the Bureau reclassified Mexican Americans as non-'white.' This reclassification was politically and economically motivated. As Ian Haney-López explains, "This classification helped legitimize federal and state expulsion campaigns between 1931 and 1935 that forced almost half a million Mexican residents—nationals and U.S. citizens alike—south across the border" ("Race" 44). In 1940, after intense lobbying by Mexican Americans and the Mexican government, U.S. citizens of Mexican descent were reclassified as "white." In the 1950s and 1960s the government continued defining them as "white" but also created a broader pan-Latino ethnic category labeled "Persons of Spanish Mother Tongue;" in the 1970s, the government underscored this ethnic dimension by renaming the category "Persons of Both Spanish Surname and Spanish Mother Tongue;"[17] and in the 1980s, these categories were replaced with the governmentally imposed term "Hispanic."

This most recent invention is extremely contentious. Many so-called Hispanics reject the term's association with Spanish ancestry and thus its 'white' eurocentric implications, as well as its erasure of their cultural specificity, and name themselves "Chicano/a," "Mexican American," "Cuban-American," and so on. Indeed, in the 1990 census over 96 percent of the 9.8 million people who refused to identify themselves according to the delineated 'races' would have been classified by the government as "Hispanic" (Fernández 143). Similarly, ten years later "97 percent of those checking 'some other race' were Hispanics"[18] (Hattam 66). As Omi and Winant observe, such changes "suggest the state's inability to 'racialize' a particular group—to institutionalize it in a politically organized racial system" (75–76). And yet, the state's efforts to racialize this very diverse population continue into the twenty-first century. According to a number of scholars, debates about the 2010 census include a government proposal that will remove the option to check "some other race," thus encouraging so-called Hispanics to self-identify as 'white.'[19] However, as Patricia Palacio Paredes notes, if this "other race" option is removed, Latinos will be classified as 'white' and consequently lose legal protections and redress for past wrongs. Thus she suggests that "Latino" should be recreated as a racial category.

The status of so-called "blacks" and "whites" is, perhaps, even more elusive and confusing. To begin with, the terms themselves are extremely inaccurate. 'White' is the color of this paper, not the color of *anyone's* skin. And people referred to as "black" could be described more accurately according to a wide range of hues, as they are in Nella Larsen's novella, *Passing*, as "ebony," "taupe," "mahogany," "bronze," "copper," "gold," "yellow," "peach," "ivory," or even "pinky white" (59). Furthermore, although many people labeled "Hispanic," "Native American," and "Asian American" have lighter skin than some so-called 'whites,' they are not classified as 'white' unless they are passing.

Though we generally think of 'white' and 'black' as permanent, tran-shistorical racial markers indicating distinct groups of people, they are not. In fact, the Puritans and other early European colonizers did not consider themselves 'white;' they identified themselves and each other as "Christian," "English," or "free," for in the seventeenth and eighteenth centuries the word 'white' did not represent a racial category. As Lerone Bennett, Jr., points out in his discussion of indentured servitude in the colonies,

Of all the improbable aspects of this situation, the oddest—to modern blacks and whites—is that white people did not seem to know that they were white. It appears from surviving evidence that the first white colonists had no concept of themselves as *white* people. The legal documents identified whites as Englishmen and/or Christians. The word *white*, with its burden of arrogance and biological pride, developed late in the century, as a direct result of slavery and the organized debasement of blacks. (40, his italics)[20]

Significantly, the colonists' early use of religious terminology to distinguish themselves ("Christians") from Native peoples and Africans ("heathens") facilitated and shaped the racialization process by implying that these human divisions were divinely ordained and therefore natural. As they separated themselves from Native and African peoples, the colonists "invest[ed] those distinctions with a divine authority that worked to naturalize race" (Zackodnik 424).[21] Racialization was economically and politically motivated. It was not until around 1680, with the shift from indentured servitude to lifelong, racialized slavery, that the term "white" was used to describe a specific group of people. As Webster explains, "[t]he idea of a homogeneous white race was adopted as a means of generating cohesion among explorers, migrants, and settlers in eighteenth-century America. Its opposite was the black race, whose nature was said to be radically different from that of the white race" (9). Yet as I will explain shortly, these seventeenth- and eighteenth-century meanings of 'white' people were quite different from today's meanings.

Significantly, then, the "white race" evolved in opposition to but simultaneously with the "black race." As peoples whose specific ethnic identities were Yoruban, Ashanti, Fon, and Dahomean were forcibly removed from their homes on the African continent and taken to the North American colonies, the Europeans adopted the terms 'white' and 'black'—with their already-existing implications of purity and impurity—and developed the concept of a superior "white race" and an inferior "black race" to justify slavery. Slavery preceded racialization. The Europeans did not originally label the people who lived in Africa "black;" nor did they see them as inhuman savages. As Abdul JanMohammed explains, "Africans were perceived in a more or less neutral and benign manner before the slave trade developed; however, once the triangular trade became established, Africans were newly characterized as the epitome of evil and barbarity" (80).[22]

The meanings of 'black' and 'white' are no more stable today than they were in the past. At various points during the twentieth century, the terms "Colored," "Negro," "black," "Afro-American," "African-American" (hyphenated), and "African American" (non-hyphenated) have all been used to describe U.S. Americans of African descent. But these terms are not synonymous; each indicates a different racial identity with specific personal, sociopolitical, and cultural implications (Gates, *Loose Canons* 131–51). And, of course, people within this racialized group self-identify in a variety ways. Thus, for example, Kimberly Jade Norwood refuses to describe herself as "African-American," explaining that

> After having met African-Americans who are White and having met Black Africans who now live in the United States and consider themselves African-Americans, I believe that the term African-American does not accurately describe me and the millions of other Black people who were born and raised in the United States and as products of the American heritage and experience. One of my favorite artists agrees: "I love being Black. I love being called Black. I love being an American. I love being a Black American, but as a Black man in this country I think it's a shame That every few years we get a change of name. . . . And, if you go to Africa in search of your race, You'll find out quick you're not an African American, You're just a Black American in Africa takin' up space."[23] (146)

Although the term 'white'—which has been used since the late seventeenth century to designate elite groups of people—seems more stable, it is not. Like other racialized categories, its meaning has undergone significant changes. Many people today considered 'white'—southern Europeans, light-skinned Jews, Irish, and Catholics of European descent, for example— were most definitely *not* 'white' in the eighteenth and nineteenth centuries.[24] Since the late 1960s, with the rise of what Stephen Steinberg calls "ethnic fever" (3), the "white race" has undergone additional changes. Once again, the redefinition of 'white' corresponded to shifts in the meaning of 'black.' As the Black Power movement developed an oppositional ideology to challenge existing definitions of "Negro," 'white' ethnics began (re)claiming their European cultural "roots." More recently, conservative self-identified 'whites' attempt to redefine themselves as the new oppressed group—people who are *disadvantaged* by their 'white'-raced skin.[25] As Omi and Winant explain, "[t]he far right attempts to develop a new white identity, to reassert the very meaning of *whiteness* which has been rendered unstable and unclear by the minority challenges of the 1960s" (117). This rearticulation of racialized identities continues today, in essays such as hooks's "Loving Blackness as Political Resistance" (in her *Black Looks*) and in demands for investigations of 'whiteness.'

Indeed, even well-meaning, progressive social scientists (who should know better) acknowledge the politically and economically motivated nature of racial formation yet analyze, document, and discuss "the black race," "the Hispanic race," "the white race," and so on as if these supposed 'races' were monolithic, God-given facts. In so doing, they reinforce oppressive social systems and erect permanent barriers between supposedly separate groups of people.[26] As Roger Brubaker, Mara Loveman, and Peter Stamatov assert,

> Despite the constructivist stance that has come to prevail in sophis-
> ticated studies of ethnicity, everyday talk, policy analysis, media
> reporting, and even much ostensibly constructivist academic writing
> about ethnicity remain informed by "groupism": by the tendency
> to take discrete, sharply differentiated, internally homogeneous,
> and externally bounded groups as basic constituents of social life,
> chief protagonists of social conflicts, and fundamental units of social
> analysis. Ethnic groups, races, and nations continue to be treated as
> things-in-the-world, as real, substantial entities with their own
> cultures, their own identities and their own interests. (45)

Like other forms of monothinking, this racialized "groupism" distorts our perception and prevents us from recognizing our interconnectedness. Rather than look for commonalities among differently situated groups, we focus only on the differences—defined in oppositional terms. Rather than work together for social change, we compete with each other.

One of the most egregious examples of this oppositional perceptual bias I have yet encountered can be found in Andrew Hacker's 1992 bestseller, *Two Nations: Black and White, Separate, Hostile, and Unequal*. In his introduction Hacker describes " 'race' " as "a human creation," not a fixed biological fact, and acknowledges that clear-cut definitions are impossible because people use the word in diverse ways (4). Yet throughout the book he continually refers to "the black race" and "the white race" without complicating the terms. Indeed, by downplaying the economic, cultural, and ethnic diversity found within each of these two supposed 'races,' Hacker heightens and solidifies the tensions between them. Moreover, by focusing almost entirely on the 'black'/'white' binary, Hacker reinforces the myth of racial purity and ignores the tremendous diversity found in this country.

Simplistic binaries between fixed definitions of 'blackness' and 'whiteness' occur in literary analyses of 'whiteness' as well. Take, for example, Nielsen's exploration of 'whiteness' in *Reading Race*. Unlike Morrison—who at least begins blurring the boundaries between 'blackness' and 'whiteness' by exploring what 'white' representations of 'blackness' tell us about literary 'whiteness'[27]—Nielsen focuses almost entirely on 'white' poets' racist stereotypes of "black" people. Although he acknowledges the fictional, contradictory nature of these 'white' representations of "blackness," his constant attention to the stereotypes themselves inadvertently strengthens the racist imagery he tries to undercut. This approach, with its focus on 'white' stereotypes of 'black' people, can be especially dangerous in our classrooms where, as Sharon Stockton points out, students already tend to think in terms of stereotyped binary oppositions. It is not enough simply to "uncover the ways that the white mind represents blackness to itself," as Nielsen hopes to do (46). We must take our analysis even further and break out of this dualistic 'black'/'white' model by encouraging students to recognize the relational nature of all representations of 'race.'

my classroom compromise

Although 'whiteness' is one of the most difficult topics to include in classroom instruction, it is crucial for educators to explore it in our teaching: We cannot analyze culturally specific dimensions of texts by writers of colors without also examining 'whiteness.' To do so further normalizes 'whiteness' and thus reinforces the long-standing hidden belief in 'white' invisibility as well as the 'white' supremacism embedded within U.S. culture. However, educators must become more aware of the complex impact investigations of 'whiteness' and 'race' can have on our students.[28] Although many self-identified students of colors find it satisfying to see the 'white' gaze that has marked them as "other" turned back on itself, I question the long-term effectiveness of this binary-oppositional reversal. As I have argued, such reversals inadvertently support already-existing stereotypes. Moreover, these oppositional maneuvers trigger a variety of unwelcome reactions in self-identified 'white' students, ranging from guilt, withdrawal, and despair to anger and the construction of an extremely celebratory racialized 'whiteness.' As Charles Gallagher explains in "White Reconstruction in the University," many contemporary 'white'-identified students now see themselves as racialized. Generally, this self-perception is accompanied by a sense of anger, frustration, and heightened alienation from people identified as non-'white.' Peter McLaren makes a similar point, noting that that many 'white'-identified students are developing reactionary, racialized identities: "Feeling that their status is now under siege, whites are now constructing their identities in reaction

to what they feel to be the 'politically correct' challenge to white privilege"
(*Revolutionary Multiculturalism* 262).

Educators must be prepared to engage with these responses. It is not
useful to demonize people who identify as "white." Nor is it helpful to
encourage feelings of self-blame for the slavery, decimation of indigenous
peoples, land theft, and other horrific events that occurred in the past.
Such highly personalized guilt is ineffective and prevents 'white'-identified
students from acting. As Ella Shohat and Robert Stam explain,

> while guilt is on one level a perfectly appropriate response to genocide,
> slavery, racism, and discrimination, it is on another level counter-
> productive. Guilt has a tendency to "curdle," to turn into a sour
> resentment toward those "provoking" the guilt. It leads to "compas-
> sion fatigue," premised on the self-flattering presumption of initial
> benevolence and the assumed luxury of a possible disengagement.
> Guilt can be narcissistic, leading to orgies of self-abuse. (344)

This solipsistic, paralyzing guilt must be transformed into an ethics of
accountability that enables students of all colors to more fully comprehend
how these oppressive systems that began in the historical past continue
misshaping contemporary conditions. Only then can they begin working
for social change.[29]

As I will demonstrate in more detail in the following chapter, the
compromise I have arrived at—admittedly temporary and always open
to further revision—entails a twofold approach where we explore the
artificial, unstable nature of 'white,' 'black,' and other racialized identities
while analyzing their concrete material effects. To begin with, I historicize
'race' by providing students with an overview of the ways racialized
identities have functioned and changed in U.S. culture. I select texts
such as those by Jesse Fauset, Langston Hughes, Adrian Piper, George
Schuyler, and Nella Larsen, which expose both racialized identities in
transitional states and the personal and cultural implications of passing. In
the stories collected in Hughes's *The Ways of White Folks*, for instance,
we see 'black' people abandoning their dark family members and reinventing themselves as 'white,' self-identified 'blacks' acting exactly like
'whites,' and racist 'white' people acting 'black.'

Because language plays such a vital role in shaping our perceptions of
reality, I pay great attention to the words I use when discussing 'race'
and racialized identities, histories, and concerns. I do not mark people,
texts, or traditions with racial labels unless the labels are context-specific
and relevant to the discussion.[30] I have replaced the superfluous references

to raced identity we are all unconsciously trained to use with more nuanced descriptors. For instance, if I am simply describing someone, I don't racialize them by referring to "the *'white'* woman," "the *Chicano*," the "*Chinese-American* boy," and so on. Instead, I select other, less loaded terms: "the woman in the red dress," "the man walking the dog," "the little boy in the green shirt," and so on. My goal here is to challenge the conventional U.S. reading practice that encourages us automatically to identify the bodies we encounter according to 'race,' defined simplistically by obvious physical differences. As I have just demonstrated, there *are* other ways to label people, when labels are necessary. (And sometimes they're not!) When referring specifically to racialized identities, I use terms like "self-identified," "raced," and "racialized;" such terms allow me to denaturalize 'race' and emphasize that racialized identities are imposed and/or chosen in a variety of ways. I refer to writers as *'white'-identified*, *'black'-identified*, *Chicano-identified*, and so on when discussing authors such as Emerson, Jean Toomer, and John Rechy, who do not emphasize their own racial identities but are labeled as such by scholars and the larger culture. When discussing writers who *do* identify themselves by their ethnic-racial affiliations I attempt to indicate this fact in my discussions, by describing them as *self-identified* "Native," "Chinese American," and so on. Through my mindful attention to language, I hope to underscore the artificial, changing nature of racialized identities even further.

To be sure, this emphasis on the particular terms we use does not in itself solve the problems I have discussed in this chapter. As David Hollinger states, "[c]hanging our vocabulary will not do much to diminish unequal treatment, but it might at least keep us aware of the direction in which antiracists want to be heading. Racism is real, but races are not" (39). Careful attention to descriptors enables me slowly to chip away at students' unconscious racialized perceptions. Selecting my words with great care, I expose and in other ways challenge racism without accidentally reinforcing the monolithic, potentially racist implications of 'race.'

In classroom discussions of 'whiteness' I no longer focus exclusively on 'white'-raced people or even on literary or cultural 'whiteness.' Instead, I define 'whiteness' as a framework—an epistemology and ethics—that operates as an invisible norm undergirding U.S. culture (the mainstream educational system, the media, and so on).[31] I do not automatically equate this epistemological-ethical 'whiteness' with all people identified as 'white.' Like Marilyn Frye, I believe that the relationship between 'whiteness' and 'white' people is contingent. But I take Frye's assertion further and suggest that in contemporary U.S. culture we are all, in various ways, inscripted into this 'white' framework. Because this framework functions to benefit 'white'-raced people, they can be—and often (although not always) are—more invested in it.[32] However, epistemological 'whiteness' is only partially about 'race,' for it intersects with certain versions of masculinity, economic status (middle- to upper-class), and colonization.

As I define the term, epistemological 'whiteness' entails an authoritative, hierarchical, restrictive mode of thinking premised on an unspoken 'white' norm. This epistemological framework operates by dividing reality into entirely separate pieces and categorizing each piece in ways that reinforce and naturalize the arbitrary divisions. Look for instance at residential segregation in the United States. As geographers Owen J. Dwyer and John Paul Jones III demonstrate, racialized segregation patterns indicate the existence of a "spatial epistemology [that] relies upon discrete categorization of space—nation, public/private and neighbourhood [sic]—which provide significant discursive resources for the cohesion and maintenance of white identities" (210).[33] Epistemological 'whiteness' transforms relational differences into a restrictive hierarchy that ranks things according to the degree of their deviation from the ('white') norm. (See appendix 4.)

Sometimes I question my insistence on so closely associating this non-relational epistemology with whiteness. Why not simply refer to it as "monothinking" or describe it as a supremacist framework? Why must I racialize it—especially given my own deep suspicions about race? I do so to acknowledge what George Lipstiz calls "the possessive investment of whiteness."[34] As my brief discussion of residential segregation indicates, historically and still today, those people categorized as 'white' have most fully benefited from this dominant epistemology and framework.

Another tactic I employ is the concept of *transcultural mestizaje*. I borrow this term from Cuban literary and political movements where its usage indicates a profound challenge to existing racial categories. As Nancy Morejón explains, transcultural mestizaje defies static notions of monolithic cultural identities by emphasizing

> the constant interaction, the transmutation between two or more cultural components with the unconscious goal of creating a third cultural entity . . . that is new and independent even though rooted in the preceding elements. Reciprocal influence is the determining factor here, for no single element superimposes itself on another; on the contrary, each one changes into the other so that both can be transformed into a third. Nothing seems immutable. (Qtd. in Lionnet 15–16)

This idea of ongoing, reciprocal transformation and change provides an important alternative to the well-known stereotype of the supposedly colorblind "American" melting-pot and the assimilationist beliefs on which it relies. Unlike the melting-pot, which turns culturally specific groups with distinct traditions into indistinguishable 'whites,' mestizaje emphasizes the mutually constituted and constantly changing nature of *all* racialized identities. Writings by Gloria Anzaldúa, Gish Jen, Chang-Rae Lee, Toshio Mori, Cecile Pineda, and Karen Tei Yamashita provide concrete examples of the many ways this transcultural process occurs.

(in)conclusion . . . a few more problems
with 'whiteness' and 'race'

But this chapter has no final solutions, no always-successful strategies for classroom explorations of 'whiteness.' What began as an investigation of 'whiteness' turned into an interrogation of 'race,' and I have even *more* concerns than I had when I began this investigation over ten years ago. I agree with Mercer and others who call for an examination of the ways 'whiteness' has been socially constructed. Because 'whiteness' has functioned as an oppressive, mythical norm that negates peoples (whatever their skin color) who do not conform to its standard—we must expose and deconstruct it. However, I still worry that this analysis simply reifies harmful status-quo stories about 'race.' As Gates explains, "we carelessly use language in such a way as to *will* this sense of *natural* difference into our formulations. To do so is to engage in a pernicious act of language, one which exacerbates the complex problem of cultural or ethnic difference, rather than to assuage or redress it" ("Writing 'Race'" 5).

As I see it, the problems with discussing 'whiteness' and other racial categories without historicizing the terms and demonstrating the relational nature of all racialized identities include (but are not limited to) the following. First, our conceptions of 'race' are scientifically and historically inaccurate; they transform arbitrary distinctions between people into immutable, "natural," God-given facts. Second, constant references to 'race' perpetuate the belief in separate peoples, monolithic identities, and stereotypes. Third, in this country racial discourse quickly degenerates into a 'black'/'white' polarization that overlooks other so-called 'races' and ignores the enormous diversity among people. And fourth, racial categories are not—and never have been—benign. Racial divisions were developed to create a hierarchy that grants privilege and power to specific groups of people while simultaneously oppressing and excluding others.

If, as Gates implies in my first epigraph to this chapter, 'race' is a text that everyone in this country unthinkingly "reads," I want to insist that we need to begin reading—and (re)writing—this text in new ways. At the very least, we should question simplistic "common-sense" conceptions of 'race'—both by exploring the many changes that have occurred in all apparently fixed racial categories and by informing students of the political, economic, and historical motivations underlying racial formation in the United States.

   In the following chapter, I discuss some of the strategies and techniques I have developed to expose and challenge students' "common-sense" readings of 'whiteness' and 'race.'

CHAPTER FOUR

Reading 'Whiteness,' Unreading 'Race'

"I feel like I'm being judged for my white skin, but I wasn't involved in the events the author describes. My grandparents immigrated to this country in the early 1900s. No one in *my* family killed Indians or had slaves!"

"I'm not white. What does this analysis of whiteness have to do with me?"

"Why are we focusing on this race stuff anyway? Why can't we talk about everything we have in common as Americans?"

I have heard these and similar comments when discussing 'whiteness' and 'race' in my classrooms. Whether they label themselves "white," "African American," "Native American," "Hispanic," "Chicano," or "American" (which for my students does not always imply 'white'), many students actively resist understanding the power, privilege, and other implications of 'whiteness.' Ironically, this denial—because it reinforces an unjust status quo and an inflexible resistance to change—is itself a crucial element of 'whiteness.' As I explained in the previous chapter, in the United States 'whiteness' generally functions as the unmarked norm—a supposedly universal category that hides its specific values, epistemology, and other attributes under the guise of a nonracialized, colorless "human nature." Its presence erased, 'whiteness' operates as the unacknowledged standard against which all so-called minorities are measured and marked. Because this invisible 'whiteness'—which is not necessarily synonymous with all 'white'-raced people—has played a pivotal role in shaping status-quo stories and in other ways naturalizing social injustice, it is *crucial* that educators investigate it, exposing its insidious power.

I am not alone in my belief that we must investigate 'whiteness' in our classrooms. In their introduction to *White Reign: Deploying Whiteness in America* Joe Kincheloe and Shirley Steinberg argue that educators must develop "a critical pedagogy of whiteness" (14). Because they believe that "[w]hite ways of being can no longer be universalized, white communication practices can no longer be viewed unproblematically as the standard, and issues of race can no longer be relegated to the domain of those who are not White," they stress the importance of "denormalizing whiteness" by exploring its racialized, non-universal characteristics (18). Similarly, in "Multiculturalism and the Postmodern Critique: Toward a Pedagogy of Resistance and Transformation," Peter McLaren emphasizes the importance of "interrogating the culture of whiteness itself." According to McLaren, "[t]his is crucial because unless we do this—unless we give white students a sense of their own identity as an emergent ethnicity— we naturalize whiteness as a cultural marker against which Otherness is defined" (214).[1]

I fully agree with these scholars and others who maintain that educators should include investigations of 'whiteness' in our teaching; however, I am troubled by the destructive impact these investigations can have on classroom dynamics. As I indicated in chapter three, I am convinced that we should not investigate 'whiteness' without carefully thinking through the implications and our goals: Why include analyses of 'whiteness' in our teaching? What do we want to accomplish by investigating 'whiteness' in our classrooms? What's the next step, once we have exposed 'whiteness' to our students? If investigations of 'whiteness' are not carried out with great care, educators risk simply reinforcing students' already-existing essentialized notions of 'race,' as well as the 'white'-supremacist frame of reference that holds these racialized identities in place.

More importantly (and perhaps more destructively), all too often the investigation of 'whiteness' turns into a "crisis" for 'white'-identified students, leading to what Michael Apple describes as "the production of retrogressive white identities" ("Foreword" ix) or what Charles Gallagher calls "whiteness . . . as an identity that evokes victimization and racist, reactionary thinking" (33). I have seen this 'white' backlash in my classrooms: When students who self-identify or are labeled as 'white' learn about recent investigations of 'whiteness,' they are compelled to recognize the insidious roles 'whiteness' plays in U.S. culture. As I explained in the previous chapter, because they often automatically associate 'whiteness' with all 'white' people, this recognition can trigger a variety of unwelcome reactions—ranging from guilt, withdrawal, and despair to anger and the construction of an extremely celebratory racialized 'whiteness.' In short, they enact a 'white' backlash which encompasses a sense of heightened alienation from people identified as non-'white' and the belief that 'white' people are victims of affirmative action and thus the newest oppressed group.[2] Although some self-identified students of colors welcome the opportunity to see the 'white' gaze that has marked them as "Other" turned back on itself, I question the long-term effectiveness of this reversal because it reinforces the status-quo story—the belief in separate 'races,' which is itself a crucial part of 'whiteness.' These reactions shut down potential agency; students of all colors view themselves as pawns in an already-existing, highly racialized system that *they cannot change*.[3]

In this chapter, I address the following questions: How can educators assist students (of all colors) in recognizing the hidden 'white' framework they have been immersed in practically since birth? How can we investigate 'whiteness' without accidentally reinforcing it? How do we expose 'whiteness' in our classrooms without triggering anger, hatred, alienation, 'white' guilt, or other forms of disavowal?

I have learned through my own teaching experiences that it is not enough simply to begin reading 'whiteness' into previously unmarked texts, for to do so contributes to this 'white' backlash and reinforces the belief in permanent, separate racial categories. Nor is it helpful to encourage 'white'-identified students to develop "a positive, proud, attractive, antiracist white identity," as some scholars suggest.[4] Nurturing positive 'white' identities in any form strikes me as quite dangerous: given the history of 'whiteness' and 'white' identity formation, even the most progressive, well-intentioned forms of 'white' pride cannot avoid reaffirming 'white' superiority.[5] As Ian F. Haney-López points out,

> The dominant racial discourse already fashions Whites as the superior opposite to non-Whites; an uncritical celebration of positive White attributes might well reinforce these established stereotypes. At the same time, because races are constructed diacritically, celebrating Whiteness arguably *requires* the denigration of Blackness. Celebrating Whiteness, even with the best of antiracist intentions, seems likely only to entrench the status quo of racial beliefs. (*White by Law* 172, his emphasis)

No matter how progressive our own intentions might be, when we affirm 'whiteness' we automatically draw on and reinforce its insidious history.

Examining, exposing, and denaturalizing 'whiteness' must be part of a larger, twofold process that includes, but goes beyond, 'white'-raced people. In addition to exposing 'whiteness,' we must develop transformative theories and practices that can divest 'whiteness.' As I define the term, 'whiteness' includes a corrosive, nonrelational epistemology and worldview that has in many ways influenced everyone in U.S. culture. Like James Baldwin, I believe that 'whiteness' represents "a moral choice," not an essential, biologically-based identity ("On Being White" 180). This "moral choice" to be 'white' is, in fact, utterly immoral: it entails an extreme lack of self-reflection; a denial of the roles 'whiteness' and 'white' people played in slavery, the genocide of Native peoples, and other forms of conquest; a refusal to recognize the interconnections among apparently different racialized groups; and dualistic thinking. As Baldwin explains, people who "think they are white . . . do not dare confront the ravage and the lie of their history. Because they think they are white, they cannot allow themselves to be tormented by the suspicion that all men are brothers [sic]" (180). I want to build on Baldwin's analysis to suggest that people of all colors can (and must!) resist 'whiteness.'[6] As I will explain in the following pages, this unmarked 'white' norm has become the framework, subtly compelling us to read ourselves, our texts, and our worlds from within a hidden 'whiteness.' We are all the products of the history of 'race,'—a history that simultaneously relies on and reinforces arbitrary divisions among people, granting privilege and power to specific groups by excluding and oppressing others. To divest ourselves of 'whiteness' we must retrieve this denied history while, simultaneously, denaturalizing all racialized identities and exposing their relational nature.

In the following pages, I discuss some of the ways I try to challenge the hidden 'white' framework and the racial identities it holds in place by employing what I call (de)racialized reading. I bracket the prefix to underscore the paradoxical, seemingly contradictory dimensions of this enterprise: On the one hand, I believe that educators must encourage students to adopt new reading practices that enable them to recognize and explore how all people and texts are racialized—marked by name, language, location, and color. But on the other hand (and simultaneously), I believe that we must describe and enact these racialized readings very carefully, in unexpected, temporarily non-racialized ways that destabilize conventional understandings of 'race' by underscoring the potentially fluid, relational nature of all racialized identities.

I focus specifically on reading for two reasons. First, as a number of scholars have noted, reading is a site of an invisible, naturalized 'whiteness.'

The unmarked 'white' norm has become the framework, subtly compelling us to read ourselves, our texts, and our worlds from within a hidden 'whiteness.' As Rebecca Aanerud asserts, "[u]nless told otherwise, the reader, positioned as white, assumes the characters are white" (37). By thus positioning readers of all colors as 'white,' conventional reading and writing practices acculturate us into 'whiteness' and train us to think and act in 'white' ways.[7] (De)racialized reading encourages students to self-reflect on this process. This self-reflection, when combined with class discussion, can expose the hidden 'white' framework. Second, I believe that reading can transform us. My life has been transformed through reading, and I've seen the lives of my students, my friends, and my colleagues transformed as well. When we read, the boundaries between writer, reader, and text become permeable; they begin breaking down. We travel into the worlds of others. Entering into their lives, we see ourselves through their eyes. Although these journeys are always partial and incomplete, they can still transform us—altering how we see ourselves and our world. Most significantly, because perception and action are closely related, these shifts in perspective can change how we act.

Contemporary science fiction writer Octavia Butler's novels have played a pivotal role in the development of my (de)racialized approach, and so I begin with a discussion of her work.[8] I then move on to apply this approach to classroom instruction where I propose four pedagogical strategies designed to reveal the hidden 'whiteness' of conventional reading practices and thus challenge student thinking and perceptions. These strategies encourage students to enact a twofold (de)racialized reading process that exposes 'whiteness' and explores the arbitrary, unstable characteristics of 'white,' 'black,' and all other racialized identities without ignoring their concrete material effects. Drawing on my own teaching experiences, I then demonstrate how (de)racialized readings can invite students of all colors both to read 'race' in new ways and to begin divesting themselves of their own 'whiteness.' This (de)racialized approach enables us to investigate 'whiteness' without reinforcing the retrogressive 'white' identities described above.

reading octavia butler's (de)racialized writing

Like the (de)racialized reading referred to in the previous section, Butler's use of 'race' to describe key characters and communities indicates a paradoxical, seemingly contradictory act: On the one hand, her characters are almost always[9] racialized—marked by name, language, location, and color (generally in this order). But on the other hand, this racialization occurs so subtly—in context-specific instances fully integrated into the narratives—that Butler seems to *de*racialize them. At times, Butler avoids racial categories almost entirely. In *Dawn*—which takes place after nuclear war has almost completely destroyed the Earth and depicts the interactions among Lilith, other Human survivors, and their rescuers, the Oankali (an ancient, extremely alien-looking species of genetic engineers)—there are only four references to Lilith's color, and none to her 'race' per se. The first description occurs at the end of chapter one; as Lilith reflects on her previous "Awakenings," she recalls when the Oankali "put a child in with her—a small boy with long, straight black hair and *smoky-brown skin, paler than her own*" (8, my emphasis). This description is context-specific and relational, triggered by her interaction with another human being.[10] Significantly, Lilith's physical appearance is described but not labeled. Those readers who do not simply skim over this description can not use it to ascertain Lilith's 'race' in any definitive sense: She could be of African, Asian, or indigenous ancestry—or perhaps some intermixture of these with European.[11]

The second reference occurs four chapters later, as Lilith leaves the only environment she has known during the past 250 years: "There had been little color in her world since her capture. Her *own skin, her blood— within the pale walls of her prison, that was all. Everything else was the same shade of white or gray*" (30, my emphasis). Like the earlier passage, this description is relational but does not allow readers to fix Lilith's 'race.' Although we know that her skin is darker than her prison's "pale walls," this information is too vague to allow us to label her by 'race.' The third reference occurs when Lilith meets Paul Titus, the first adult human she has encountered in over two centuries: "She stared at him. A human being—tall, stocky, *as dark as she was*, clean shaved" (89, my emphasis). Once again, Lilith's physical appearance is described but not labeled, and again the description is relational. The final reference occurs almost 100 pages later, when Lilith's ooloi mate, Nikanj, commenting on her alliance with Joseph Shing, states that although other Oankali believed she " 'would choose *one of the big dark ones because they're like you*. I said you would choose this one—because he's like you. . . . During his testing, his responses were closer to yours than anyone else I'm aware of. He doesn't look like you, but he's like you' " (171, my emphasis). As in the previous examples, the characters are described but not labeled or grouped according to 'race.' In this passage, and in Butler's work as a whole, affinity between humans is based on ways of thinking and acting, not on appearance, ancestry, or "blood."

Even when Butler does label her characters' 'race' more overtly, the labels often appear unexpectedly. In *Parable of the Sower*, which depicts Lauren Olamina as she attempts to survive and create a new community in a radically altered United States, Butler does not immediately inform us of her characters' racialized identities.[12] The novel begins in the year 2024, but it is not until the following year, after readers have become acquainted with Lauren, her family, and her neighbors, that we read the following passage:

> We ran into a pack of feral dogs today. We went to the hills today for target practice—me, my father, Joanne Garfield, her cousin and boyfriend Harold—Harry—Balter, my boyfriend Curtis Talcott, his brother Michael, Aura Moss and her brother Peter. Our other adult Guardian was Joanne's father Jay. He's a good guy and a good shot. Dad likes to work with him, although sometimes there are problems. *The Garfields and Balters are white, and the rest of us are black.* That can be dangerous these days. On the street, people are expected to fear and hate everyone but their own kind. . . . Our neighborhood is too small for us to play those kinds of games. (31, my emphasis)

Even within this paragraph—which occurs in the third chapter, on the novel's thirty-first page—Butler subtly and slowly racializes her characters, and once again this racialization is context-specific and relevant to the plot.[13] Significantly, the focus is not on 'race' itself but rather on the implications 'race' makes for the characters' survival. And as in *Dawn*, alliances are not based on simplistic racial designations but instead have their source in characters' thoughts, actions, values, and beliefs.

Kindred provides an even more remarkable example of this delayed racialization. Set simultaneously in 1976 California and in pre–Civil War Maryland, this novel follows Dana Franklin and her husband, Kevin, as Dana is pulled back through time whenever the life of her ancestor, Rufus, is threatened. The 'white' son of a slaveholder, Rufus represents a shocking, previously hidden element of Dana's family heritage which she must learn to understand and accept. Significantly, it is not until the third chapter, and Dana's second trip into the past, that Butler identifies Dana as "black." Only thirty pages later does she label Kevin "white."[14] As in Butler's other novels, the descriptions are context-specific and relational. We learn that Dana is "black" only when Rufus refers to her in derogatory terms, and we learn that Kevin is 'white' only when, employing a recursive narrative style, Butler describes Dana's first encounter with this man she would later marry.[15]

What should we make of these (de)racializing tactics? Do they simply encourage some (perhaps 'white'-identified) readers to ignore the important role 'race' plays in Butler's texts? Do readers caught up in Butler's fast-paced narratives (unconsciously) assume that, because the characters are not overtly racialized, they have no 'race'—in other words, that they are 'white'? After all, as I mentioned above, U.S. readers (of all colors) have generally been trained to view 'whiteness' as the unmarked, nonracialized norm. As Toni Morrison asserts, "To identify someone as a South African is to say very little; we need the adjective 'white' or 'black' or 'colored' to make our meaning clear. In this country it is quite the reverse. American means white, and Africanist people struggled to make the term applicable to themselves with ethnicity and hyphen after hyphen after hyphen after hyphen" (*Playing* 47).

I would suggest, however, that Butler enacts a different strategy. She foregoes the hyphens (and often even the word "American" itself) and depicts her characters first as people, marked by language, location, and color—but not racial labels. Only later, when readers have already accepted the characters and entered into their lives, does Butler indicate color, and she does so contextually—in ways that are relevant to the plot. By so doing, she compels readers to recognize both the relational, contingent nature of 'race' and its profound social effects.

Butler's (de)racializing practice can challenge us to read 'race' in new ways. As such, it offers an important alternative to the status-quo stories about 'race' circulating in contemporary U.S. culture, stories that conflate color with 'race' and divide people into distinct groups based on apparent (physical) differences. As David Shipler notes in his study of U.S. 'black'/'white' interethnic relations, in today's highly racialized culture "there is no more potent attribute than the color of the skin" (232). We have been trained automatically to classify each other according to racialized appearances: "Color is the first contact between blacks and whites [and many other racialized people as well]. It comes as the initial introduction, before a handshake or a word, before a name, an accent, an idea, a place in the hierarchy of class, or a glimpse of personality. . . . From across a room or across a street, from a magazine page or a television screen, it is this most superficial attribute that suggests the most profound qualities" (231).[16] Put differently, we have internalized a 'white' reading practice that instructs us to read the bodies we encounter according to 'race,' defined simplistically by obvious physical differences. Marking her characters in unobtrusive, context-specific ways, Butler interrupts this color-based mode of reading and the status-quo stories it simultaneously relies on and reinforces. Her (de)racialization can transform readers' perceptions and challenge us to self-reflect. Without clearly defined racial differences, her characters cannot be labeled and categorized by 'race.' As the boundaries break down, inviting readers to reexamine our preconceptions about racialized identities, we begin recognizing that affinity—and, by extension, alliances among people—cannot be reduced to 'race' or to other social identity categories but instead depends on the choices we make.

Take, for example, the transformations in perception that often occur while reading *Kindred*. For the past fifteen years, I have taught this novel in a variety of courses, ranging from introductory-level women's studies and composition to graduate-level literature and women's studies courses. I have repeatedly seen my students' thinking altered through their reading and discussion of this text. No matter how they self-identify—whether as "Hispanic," "Chicana," "Indian," "black," "American," or 'white'—students' reactions have unfolded according to the following pattern. At the outset, students reading this novel do not even think about Dana's ethnicity/'race' or, if they do, they assume she is 'white.'[17] When they discover that Rufus is the 'white' progenitor of Dana's 'black' family tree, they experience several shifts in perception: First, they must confront their unconscious assumption that Dana (because unmarked) is 'white.' This confrontation exposes both the limitations in their belief that in today's "post-Civil-Rights" United States, we live in a color-blind society (a belief which many students hold dear) and the insidious ways 'whiteness' serves as the unspoken *racialized* framework. Their own reading experience has just demonstrated that contemporary U.S. culture is not as color-blind as they generally believe. They have *not* read the unmarked Dana as an un-raced individual but instead have unconsciously read her as 'white,' vividly illustrating how 'whiteness' functions as the invisible, supposedly raceless norm. Through class discussion, students can recognize some of the ways that this hidden 'whiteness' informs contemporary beliefs about the supposedly color-blind laws, thus reinforcing existing injustices.

Second, and like Dana herself, most students are quite startled by the implications of her newly discovered 'white' relative. The fact that Dana was unaware of her own 'white' ancestor reveals the flawed nature of U.S. status-quo stories about 'race'—stories that rely on facile concepts of racial purity. People we identify as 'black' cannot be so neatly categorized into a single 'race.' Indeed, as Naomi Zack suggests, "It has been estimated that between 70 and 80 percent of all designated black

Americans have some degree of white ancestry" (*Race* 75). And the fact that Rufus has two mixed-'race' children, one of whom looks 'white,' indicates that many people whom we read/define as 'white' also have unknown mixed ancestry. A third, closely related transformation occurs when, having then assumed that Kevin must also be 'black'—simply because he is Dana's husband—students learn thirty pages later that he is not. Again, they must reexamine their racialized preconceptions and the status-quo stories they have internalized and unconsciously act out as they read, for alliances cannot be ascertained by focusing solely on 'race.'

As my students' experiences demonstrate, reading can transform us. Françoise Lionnet makes a similar point: "one does not enter into a fictional world without risk, the risk of being influenced by a specific point of view. Reading is a two-way street and by implicating myself in my reading, I am in turn transformed by that activity" (28). When readers immerse themselves in the fictional worlds of Butler's novels, our usual highly racialized ways of viewing the world are challenged. We risk being changed by the (de)racialized characters we encounter. Butler's textual revelations and the self-reflection they provoke can compel readers to acknowledge and explore both our own assumptions concerning racial purity and the ways we label, categorize, and stereotype people according to 'race.' Despite their commonly held assumptions that "we" have moved beyond 'race' and judge people simply as individuals, students' surprise at discovering Kevin's and Dana's racialized identities helps them to understand that in fact color and 'race' shape our perceptions in unacknowledged, sometimes dangerous ways. We do not read the bodies we encounter simply as unmarked human beings; they are 'raced'—even when that 'race' is (unmarked) 'white.' Although it seems paradoxical, Butler's (de)racializing practice challenges readers to recognize that those bodies which *seem* unmarked are actually marked 'white.'

(de)racialized teaching strategies

I have borrowed this (de)racializing process to develop four strategies which I now use in various ways when teaching about 'whiteness' and 'race.' Without seeming to focus on 'whiteness' or, more generally, on 'race,' these strategies—which encompass mindful text selection, delayed racialization, denaturalized racialization, and the deconstruction of 'whiteness'—enable me to expose 'whiteness' while de-essentializing *all* racialized identities. After outlining these pedagogical strategies, I will explain how I applied them in an introductory literature course.

Although I don't expect that you can take these strategies and automatically use them in your own teaching, I do hope that you might be able to adapt them in ways suitable to your specific situations.

First, I select texts carefully, and think through what I want students to learn about 'race' as they read. As I explained earlier in *Teaching Transformation*, like the larger U.S. culture in which they live, the majority of my students assume that 'race' is an unchanging biological (and divine) fact, based on natural (God-given) divisions among people. They believe that each individual is born into a single 'race,' defined monolithically, which can be read simply by looking at physical appearance. When teaching about 'whiteness' and 'race,' I want to challenge this naturalized taxonomy of 'race' and invite students to explore how the belief in discrete, biologically separate 'races' relies on nineteenth-century pseudoscientific theories that have been disproved.[18] Not surprisingly, my students' beliefs about 'whiteness'—when they have thought about it at all—are equally misguided. Because 'whiteness' has functioned as the unmarked norm, students have rarely (if ever) explored its racialized epistemological dimensions. I want students to recognize its pervasive yet invisible nature, as well as the power dynamics that underlie and reinforce 'whiteness' and all other racialized identities. However, I do not want them to insert 'whiteness' into their already-existing status-quo stories and then view 'white'-raced people as just one more homogeneous 'race.' I try to select texts enabling me to explode racial stereotypes, texts demonstrating that there is no monolithic definition of 'whiteness' or of 'white' people, just as there is no monolithic definition of 'blackness' and 'black' people, 'Asianness' and 'Asian American' people, or any other racialized group. To assume otherwise simply reinforces stereotypes and prevents us from recognizing potential commonalities.

Second, I delay racialization. I rarely begin a course or a unit by focusing specifically on 'race' or on any other social identity category; instead, I start with thematic and aesthetic issues and give students specific suggestions, topics to reflect on as they read. I have learned that these topics should not foreground 'race,' for to do so often makes students defensive and triggers various status-quo stories. I plan classroom discussion carefully, and lead students through questions designed to assist us in becoming aware of how they have unconsciously operated within a 'white' reading framework. As I will explain later, I find it effective to begin with seemingly "universal" topics (such as quest motifs, relationships, childhood, and so forth), rather than with 'race' or any other social identity category.

Third, I denaturalize 'race.' When I (finally) introduce the concept of 'race' into our discussion, I present it in ways that historicize and pluralize racial categories. Offering students an overview of some of the ways racialized identities have originated, functioned, and changed in U.S. culture enables me to demonstrate the artificiality in the racial stories we have all been trained to read. I give students a short historical lecture discussing the relational, economically driven origins of U.S. racial categories. My goal here is twofold: (1) I want to reveal the limitations in the status-quo stories students employ when they read—stories that present 'race' as permanent, monolithic components of each person's identity; and (2) I want to pluralize 'race' by emphasizing that racialized identities have changed over time and take diverse forms.

Fourth, after exposing the insidious components of 'whiteness,' I invite students to deconstruct it by exploring the 'white' frames of reference that in many ways shape their worldviews. This final step has the potential to give students agency and calls them to accountability. As students learn to read 'whiteness' in previously unmarked bodies and texts, they can recognize the political implications of 'race' and begin deconstructing the status-quo stories they have been trained to read.

These four (de)racialized reading strategies encompass a subtle movement designed to explore 'whiteness' yet make 'race' more unstable, and to do so without overlooking the racism, 'white' supremacy, homophobia, and other forms of individual and systemic social injustice. After intentionally *de*racializing student reading practices and classroom discussions, I *re*-racialize them through close textual analyses of specific passages and discussions focusing on 'whiteness.' These passages demonstrate that 'race' is an artificial, relational meaning system that can (and, in my opinion, must!) change. By thus emphasizing the constructed, historically changing nature of 'race' while exposing 'whiteness,' (de)racialized reading strategies make students accountable and invite them to unread 'race' by reading it in new ways. I now want to illustrate one form these (de)racialized reading strategies can take in the classroom. In the following pages I describe a unit on 'whiteness' and 'white' studies I designed and taught in Introduction to Literature, a two-hundred-level course required for English majors and a popular general education elective for non-majors.

(de)racializing 'whiteness'

1. *Selecting Texts.* I selected two texts for this unit: Don DeLillo's 1985 novel, *White Noise*, and Leslie Marmon Silko's 1977 novel, *Ceremony*. I chose the former because, despite its title, it seems to have nothing to do with 'race' yet (as I will explain later) contains highly racialized subtexts,

including provocative references to 'whiteness;' and I selected the latter because it challenges and denaturalizes restrictive definitions of 'whiteness.' Like *White Noise*, *Ceremony* is filled with images of 'whiteness' and 'white' people. But unlike *White Noise*, *Ceremony* openly de-normalizes 'whiteness' by exposing it and putting it in dialogue with other racialized groups. Despite the many differences between them, the novels share a number of similarities. Both focus primarily on a single figure: In *White Noise* the protagonist is Jack Gladney—an upper-middle-class, highly educated man, creator of "Hitler studies" as an academic discipline, the chair of "Hitler studies" at College-on-the-Hill (a small liberal arts college somewhere in the United States) who, we later learn, is 'white;' and in *Ceremony* the protagonist is Tayo, a light-skinned mixed-blood World War II veteran suffering from post-traumatic shock, who feels personally responsible for the deaths of his uncle and cousin as well as the drought devastating New Mexico. Each protagonist experiences a crisis in his identity: In *White Noise*, Jack is obsessed and paralyzed by his fear of death, a fear intensified by his perhaps deadly exposure to environmental waste during the "airborne toxic event;" and in *Ceremony* Tayo returns from the war alienated both from the larger dominant culture and from traditional Laguna Pueblo culture. As they attempt to make sense of their lives, each protagonist enacts a type of quest. Throughout much of *White Noise* Jack's quest is internal and self-analytical: During conversations and meals with family members and colleagues and excursions to the supermarket and shopping mall, he attempts to master his fear of death; in the novel's final section Jack's quest takes a more outward form as he searches for Willie Monk, creator of Dylar, an experimental drug designed to cure people's fear of death. Tayo's quest is more visionary and mythic: Through a series of ceremonies and rituals with Ku'oosh, the Pueblo medicine man; Betonie, the mixed-blood/Navajo healer; and Ts'eh Montaño, a mythic-yet-embodied supernatural woman, Tayo attempts to understand his role in the cosmos; his quest culminates in a final conflict with his fullblood childhood friends, Emo, Harley, and Leroy, who have surrendered themselves to the witchery and the seductive 'white' culture that Tayo must learn to resist.

2. *Delaying Racialization.* Before beginning this unit on 'whiteness,'[19] I told students that we would focus on issues concerning character, identity formation, and quest motifs. In the syllabus, I included a list of recommended questions for students to think about while reading the novels:

> How would you describe Jack and Tayo? How do they view themselves? What similarities and differences do you find between

them, and how do you account for these similarities and differences? How does each man attempt to construct his identity? What roles do other people and the larger dominant culture play in each protagonist's identity construction? How do Jack and Tayo change in the course of the novels? What specific crises do Jack and Tayo face? How do they respond to these crises, and how do you account for their responses? Do the texts use this crisis-and-response to comment on contemporary social issues and, if so, how?

For student readers, these questions seem to be totally unrelated to 'whiteness' or more generally to 'race.' They simply focus on some of the most obvious textual issues that might arise during a first reading of the novels.

I began with *White Noise* because it allowed me to introduce and (de)racialize the topic of 'whiteness,' a topic *Ceremony* picks up on in more overt ways. During the first two days of this unit, we discussed the issues referred to above. Significantly, no one (including me) mentioned 'race.' This omission on my part was intentional, and designed to mimic Butler's (de)racializing strategies. This omission on my students' part was unintentional and reflects the fact that they have been trained to ignore 'whiteness.' By introducing 'whiteness' and 'race' in a larger context— that of each protagonist's identity crisis—I hoped to assist students in recognizing the enormous roles 'whiteness' and 'race' play in U.S. culture without reinforcing their already-existing racial scripts.

On the third day of this unit, I finally introduced the topics of 'whiteness' and 'race.' Even at this point, I did not focus specifically on 'whiteness' or 'race' but instead situated these concepts in a larger framework—that of literary and cultural studies as a whole. In so doing, I hoped to challenge student scepticism about 'whiteness' and prevent them from falling back on retrogressive essentialized 'white' identities. I began class by announcing that 'whiteness' studies was the topic of the day, and I explained *why* we would examine this topic, a topic that most of my students had never considered. Because 'whiteness' studies is a growing field in cultural and literary studies, it is important that they be familiar with it. As I gave students background information on the development of this interdisciplinary field, I repeatedly underscored its relevance to the study of literature. I explained that for several decades scholars have looked at the racialized dimensions of literary works by authors identified as "African-Americans," "Native Americans," "Chicanos/as," and so on, but they have not examined the racialized dimensions of works by authors identified as 'white.' I offered concrete examples to reinforce my point: Although the

Harlem Renaissance is generally described as a "*black*" literary movement, scholars do not describe Puritanism or Transcendentalism as "*white*" movements, even though—to the best of my knowledge—the Puritans and Transcendentalists were all people of European descent. There are studies of "*Chicano*" narrative, "*Asian-American*" novels, "*Native American*" poetry, and so on. But, I asked students, how often do you see courses or books devoted exclusively to 'white' writers—as so many courses and books still are—that acknowledge this fact in their titles: Say, "Classics of the *White* Western World," "The *White* American Experience," or "*White* Regional Writers"? In this schema, I explained, "minority" writings become deviations from the unmarked, unspoken 'white' norm.

By focusing on course titles and literary categories in this way, I (de)racialized them and invited students to think consciously about how we are trained to read and classify literature. I bracket the prefix to emphasize my contradictory approach: I racialized the categories yet did so in a manner that encouraged students to recognize their artificial, constructed nature. This approach allowed me to demonstrate three concepts: the highly racialized character of literary categories; the fact that, as readers, we are too often unaware of this racialization; and the hidden 'whiteness' of canonical U.S. texts.

3. *Denaturalizing 'Race.'* This third step, denaturalizing 'race,' involved three interrelated strategies: (1) exploring the U.S. origins of 'whiteness;' (2) historicizing 'race;' and (3) exposing the fluidity and multiplicity in racial designations. These strategies enabled me to challenge students to reflect on their perceptions. I invited them to question and perhaps reevaluate the racialized status-quo stories they have been trained to read and act out. As I indicated in the previous chapter, we should not talk about 'whiteness' without also talking about 'race.' The trick, however, is to discuss these issues without reinforcing status-quo stories and other essentializing concepts. To do so, I offered students an overview of the development of racialized identities in the United States, with special emphasis on the constructed, relational nature of 'black' and 'white.' I wanted students to recognize that although people in this country generally view these terms as permanent, ahistorical racial markers indicating genetically distinct groups of people, they are not. Like other social identity categories, these so-called 'races' were shaped by economic and political concerns; they functioned materially and psychically—humanizing and privileging some people while dehumanizing, segregating, and in additional ways violating others.

I began denaturalizing 'race' by talking about 'white'-raced identity. I emphasized the fact that the Puritans and other early European colonizers did not consider themselves 'white,' for in the early seventeenth century the word 'white' did not represent a racial category. Indeed, before the invention of a 'white' racial category, poor and enslaved people of various colors/ethnicities had a great deal in common. As Lerone Bennett, Jr., notes, in the seventeenth century indentured servants created a genuinely color-blind community:

> Working together in the same fields, sharing the same huts, the same situation, and the same grievances, the first black and white Americans, aristocrats excepted, developed strong bonds of sympathy and mutuality. They ran away together, played together and revolted together. They mated and married, siring a sizeable mixed population. In the process the black and white servants—the majority of the colonial population—created a racial wonderland that seems somehow un-American in its lack of obsession about race and color. There was, to be sure, prejudice then, but it was largely English class prejudice which was distributed without regard to race, creed, or color. (40)

I discussed these color-blind communities in some detail because I want students to recognize that alternative communities are possible. If, in the past, we have made alliances based on affinities and life experiences rather than on 'race,' we can do so again. I hope that these historical examples will liberate students' imaginations, enabling them to envision and enact additional possibilities.

I underscored the fact that racialization was economically and politically motivated: It was not until around 1680, with the racialization of slavery, that the term 'white' was used to describe a specific group of people, a group that excluded many people today considered 'white.'[20] I drew heavily on Thandeka's work as I discussed the close relationship between the racialization of slavery and the development of a 'white' race. As Thandeka explains in her analysis of colonial Virginia's late-seventeenth- and early-eighteenth-century laws, "Virginia's race laws . . . marked the beginning of a new era during which the camaraderie among persons of different colors who found common cause both because of similar social circumstances and class status gradually gave way to a racial antipathy that put an end to any interracial advance toward common class interests" (47). This racial animosity was not automatic; some newly 'white'-raced people resisted this division.[21] By emphasizing that not all 'white'-raced people participated in this early racism, I hope to tease apart automatic equations between 'white' identity and racism.

I want my self-identified 'white' students to realize that they are not inherently (by virtue of their so-called 'race') racist. They have agency. If they are willing to do the difficult work of self-reflecting on the various forms racism takes in their environments and lives, they can actively work to transform it by thinking and acting in nonracist ways.

I underscored the relational nature of racialized categories by explaining that the "white race" evolved in opposition to but simultaneously with the "black race." It was not until slavery became racialized that the Europeans adopted the terms 'white' and 'black'—with their already existing implications—and developed the concept of a superior "white race" and an inferior "black race" to justify slavery. I used the history of "passing" to destabilize this binary between 'blackness' and 'whiteness' even further. Drawing on my personal experience, I briefly informed students about my own family background—my grandmothers who tried to pass into 'whiteness' and my family's attempts to forget and in other ways deny the implications of our ancestry. I emphasized that my family's experience is by no means unique. During the past 300 years, many thousands of people "passed" from 'blackness' into 'whiteness.'[22]

4. *Deconstructing 'Whiteness.'* After this extensive background discussion, I explained my rationale for examining 'whiteness' in *White Noise*: The title's reference to an apparently nonracialized 'whiteness' invites us to think further about what it means to be 'white;' and I lead students through a close reading of specific passages that illustrate DeLillo's selective racialization, where only non-'white' or racially ambiguous people are marked and thus denormalized: the "woman who lived on our street with a teenage daughter and an *Asian* baby" (39, my emphasis), " 'the *black* girl who's staying with the Stovers' " (80, my emphasis), the "*black* family of Jehovah's witnesses" Jack meets during the evacuation (132, my emphasis)—to mention only a few of the many instances where this selective racialization occurs.[23] In these passages—as in the novel as a whole—Jack, his family, the Stovers and other ('white') neighbors are not marked by color or 'race.' I used this textual evidence to demonstrate how an invisible 'white' framework and the selective racialization it supports confer normalcy on the unmarked (and seemingly non-'raced') characters while singling out non-'whites' as different—deviations outside the ('white') norm. As I had suspected, my students had not noticed this selective racialization when reading *White Noise* because it paralleled their own reading habits. And so, as we closely examined this textual evidence, they were forced to reflect on their own reading practices. They too have internalized this invisible 'white' norm, a norm that makes it seem natural to read 'race' selectively.

After discussing the hidden allusions to 'whiteness' in *White Noise*, I reminded students that 'whiteness' becomes more visible in the novel's final section, where we read that Jack himself is marked as "white." We read that he feels 'white' (242–43) and that he walks on his "bare *white* feet" (244, my emphasis). During his encounter with Willie Mink, Mink refers to Jack several times as "white," first identifying him as " 'a heavyset *white* man' " (308, my emphasis), later calling him " '*white* man' " and telling him that he is " 'very *white*' " (310, my emphasis). Finally, the room where this confrontation occurs is itself a "*white* room" filled with a "*white* buzz" (312, my emphasis). I suggested that this sudden proliferation of references to 'whiteness'—which occurs when Jack's fear of death and, relatedly, his inability to control or make sense of his life grow even stronger—almost begs us to explore 'whiteness' in *White Noise*.

At this point in class discussion, most students were convinced of the value in exploring 'whiteness' in *White Noise* and were willing to examine it in greater detail. I then discussed traits often associated with 'whiteness,' distributed a handout, "Critics on Literary and Cultural Representations of 'Whiteness' " (see appendix 5), and invited students to apply these 'white' traits to *White Noise*. Specifically, we explored (first in small groups and then with the class as a whole) the following questions:

- In what ways does the novel substantiate, challenge, and/or revise scholars' descriptions of 'whiteness'?
- How does incorporating an analysis of 'whiteness' affect your interpretation of the novel?
- What do you make of the associations between Jack's consumerism, his fear of death, and his 'whiteness'?

We used the handout again one week later when discussing Silko's representations of 'whiteness' in *Ceremony*. On both occasions, I tried to shape class discussions in ways designed to underscore students' intellectual agency. I emphasized that I did not want the class to automatically accept the theorists' analyses of 'whiteness.' Instead, I wanted them to engage in a transcultural dialogue of sorts with the scholarship, the literary texts, and their classmates. Through this dialogue, they should arrive at their own insights. I presented this information about 'whiteness' as hypothetical and explained that, as we talk about 'whiteness' in the assigned texts, we would not try to arrive at a definitive understanding of the concept. Instead, we would simply speculate on possible, temporary meanings of 'whiteness' and the potential usefulness of examining 'whiteness' in *White Noise* and *Ceremony*. My careful, tentative class discussion was motivated by three goals. First, by presenting 'whiteness' in this open-ended, nondogmatic fashion I hoped to prevent students from responding in defensive, reactionary ways. Second, by discussing the roles theorists and social scientists play in defining 'whiteness' and other racial categories, I wanted students to recognize the various ways knowledge is constructed. Third, and closely related, I wanted to encourage students to view 'race' as impermanent and open to change.

Class discussions of 'whiteness' in *White Noise* and *Ceremony* were truly remarkable. Students gained new insights about previously puzzling dimensions of each text, including Jack's fear of death and the ways his faith in science, technology, and consumption impede his development and interfere with self-understanding; and Tayo's alienation, his sense of inferiority, and the dominant U.S. culture's multilayered, systemic oppression of Native cultures. Perhaps not surprisingly, after examining 'whiteness' in *White Noise* my students viewed 'whiteness' and 'white' people in highly critical ways, and determined that 'whiteness' is extremely destructive. 'Whiteness' gave Jack an unspoken, unacknowledged sense of superiority and the unwarranted belief that he should be able to fully control his own life. Students read Jack's fear of death and his role as creator of Hitler studies as an indictment of 'whiteness' and an implicit equation of 'whiteness' with mass murder and other forms of death. In addition to concluding that Jack becomes paralyzed by his unacknowledged 'whiteness,' they attributed his feelings of emptiness, his fear of death, his attraction to the supermarket and shopping mall, and his apparently meaningless life to his 'whiteness.' Several students decided that 'white' spirituality is an enormous vacuum or void, an anti-spirituality of sorts that promises fulfillment through a misguided (and perhaps deadly) faith in science, technology, and consumption of material goods.

I am very aware of the ways these depictions demonize 'whiteness.' As an educator, I want students of all colors to recognize the racialized nature of 'whiteness.' However, I do not want my 'white'-identified students to assume that, because they are 'white,' these characteristics must automatically, inevitably apply to them. I do not want them to be filled with paralyzing guilt if they see in Jack's unspoken 'white' superiority and privilege reflections of their own. (As I have asserted earlier, such highly personalized guilt is ineffective and prevents 'white'-identified students from acting.) Nor do I want them to be filled with anger at these negative representations of 'whiteness' and resentfully declare themselves the most recent oppressed group. Relatedly, I do not want to foster antipathy toward all 'white'-raced people in students who do not identify as 'white.' Nor do I want them to grow smug in their own non-'white' racialized identities. My goals are quite different. As I explain in the following section, I want to assist students of all colors in developing an ethics of accountability, which enables them more fully to comprehend how these oppressive racialized systems that began in the historical past continue misshaping contemporary conditions. Only then can they form new alliances that go beyond existing racial categories and empower them to work for social change.

But in order to do so, students must recognize the constructed, contingent nature of 'whiteness' and, more generally, of 'race' as a whole. As I explained earlier, 'whiteness' is not synonymous with 'white' people. The fact that a person is born with 'white' skin does not automatically ensure that s/he will think, act, and write in the 'white' ways described in this and the previous chapter. Nor does the fact that a person has "brown" or "black" skin automatically ensure that s/he will *not* think, act, and write in 'white' ways. As Alison Bailey points out,

> The connection between "acting white" and "looking white" is contingent, so it is possible for persons who are not classified as white to perform in whitely ways and for persons who are white not to perform in whitely ways. Racial scripts are internalized at an early age to the point where they are embedded almost to invisibility in our language, bodily reactions, feelings, behaviors, and judgments. (34)

It is crucial for educators to assist students in recognizing the pivotal role 'whiteness' plays in supporting the racial status-quo stories we have all been trained to read and enact. In the United States, 'whiteness' serves as the unacknowledged framework, a framework that has affected us all. No matter how we identify—whether as "Chicana," "Hispanic," "Mexican," "Anglo," "American," "black," "white," "multiracial," or so forth—we all, to greater and lesser degrees, have learned to think, read, and act in 'white' ways. We have internalized what Gloria Anzaldúa describes as a 'white' "frame of reference." As she explains in a 1996 interview with Andrea Lunsford, in the United States

> the frame of reference is white, Euro-American. . . . [W]e—the colonized, the Chicanos, the blacks, the Natives in this country—have been reared in this frame of reference, in this field. All of our education, all of our ideas come from this frame of reference. We're complicitous because we're in such close proximity and intimacy with the other. . . . This white culture has been internalized

in my head. I have a white man and woman in here, and they have me in their heads, even if it's just a guilty little nudge sometimes. (*Interviews/Entrevistas* 252–54)

Krista Ratcliffe makes a similar point. Distinguishing between 'white' people and 'whiteness,' she asserts that "[l]ike any other socially constructed category (student, teacher, dean, gender, race, class) whiteness is a trope, and the actions and attitudes associated with this trope are embodied in all of us (albeit differently) via our socialization" (110–11). Unlike Ratcliffe and Anzaldúa, however, many students are almost entirely unaware of this 'white' framework until we invite them to explore its pervasiveness in their lives.

Although I mentioned this 'white' framework during class discussions of *White Noise*, it was not until students read *Ceremony* that could they recognize the contingent nature of 'whiteness' and skin color: The full-blood Native characters such as Emo, Harley, and Rocky think and act in 'white' ways, whereas the light-skinned, part-'white,' mixed-blood protagonist, Tayo, learns to recognize and resist 'whiteness.' To be sure, Silko in many ways demonizes 'whiteness': in *Ceremony* the dominant 'white' culture is associated with greed, restrictive boundaries, destruction, emptiness, absence, and death. However, she does not essentialize 'whiteness' by placing it only within specific, racially marked bodies. Instead, she gives us a story—an origin myth, told by Betonie, the mixed-blood Navajo healer. In response to Tayo's statement " 'I wonder what good Indian ceremonies can do against the sickness which comes from their ['white' people's] war, their bombs, their lies,' " Betonie asserts:

> "That is the trickery of the witchcraft. . . . They want us to believe all evil resides with white people. Then we will look no further to see what is really happening. They want us to separate ourselves from white people, to be ignorant and helpless as we watch our own destruction. But white people are only tools that the witchery manipulates; and I tell you, we can deal with white people, with their machines and their beliefs. We can because we invented white people; it was Indian witchery that made white people in the first place." (132)

In this passage, and in *Ceremony* as a whole, 'whiteness' is depicted as a highly destructive, immensely seductive though sadistic worldview that—with great effort—can be resisted. It is *not* European ancestry that dooms 'white' people to act in these evil ways but rather a system of values and a code of conduct potentially adopted by people of any color. Like Baldwin in "On Being White," Silko represents 'whiteness' as "a moral choice" (180) rather than an essential, biologically-based identity. As Betonie's creation story and Tayo's quest demonstrate, once we begin recognizing this previously invisible 'whiteness,' we can—if we so desire—resist it.

Through class discussion, I emphasized the relationship between agency, reading, and self-reflection; together, they can challenge 'whiteness.' More specifically, I invited students to acknowledge yet *dis*-identify with the representations of 'whiteness' they encounter in *White Noise* and *Ceremony*. As Anzaldúa explains in "To(o) Queer the Writer," "Reading is one way of constructing identity. When one reads something that one is familiar with, one attaches to that familiarity, and the rest of the text, what remains hidden, is not perceived. Even if one notices things that are very different from oneself, that difference is used to form identity by negation—'I'm not that, I'm different from that character. This is me, that's you' " (257). (De)racialized readings can trigger this identification/dis–identification process. Before class discussion, students read 'race' unthinkingly, thus reinforcing the 'white' frame of reference holding racialized identities in place. They automatically, unthinkingly inserted themselves into the status-quo stories they have been trained to read. By exposing these stories—and inviting students to consciously reflect on what they had unconsciously assumed—I opened space for dis-identification to occur. Dis-identifying themselves with the representations of 'whiteness' they encountered in *White Noise* and *Ceremony*, they could begin divesting themselves of 'whiteness.'

This process of dis-identification was especially useful for 'white'-identified students. Although it sounds paradoxical, these students became more accountable for their own 'white'-raced identities yet simultaneously recognized that they did not need to align themselves with Jack Gladney or the 'white' people in *Ceremony*. They realized that their racialized 'whiteness' does not force them to act in these negative 'white' ways. As in Butler's novels, affinity is based on ways of thinking and acting, not (necessarily) on 'race.' They can dis-identify themselves with the representations of 'whiteness' in *White Noise* and *Ceremony*, and begin developing new alliances and actions based on ways of thinking and acting, not on 'race.' This recognition made them answerable yet prevented them from falling into "a personalistic, neurotic guilt" (Shohat and Stam 343–44).

Because reading 'whiteness' into apparently nonracialized texts allows us to begin disrupting 'whiteness' as the unspoken, unquestioned norm,[24] the (de)racialized reading strategies I employed in this unit on 'whiteness' were extremely effective. I was able to investigate and expose 'whiteness' and its negative power without triggering a crisis in 'white'-identified students or in other ways contributing to a 'white' backlash. Students (of whatever colors) could no longer ignore the insidious role 'whiteness' plays in normalizing our hierarchical social system. Course texts and class discussions vividly demonstrated that—popular rhetoric to the contrary—contemporary U.S. culture is *not* color-blind. We have all, to varying degrees, been socialized into an invisible, highly destructive 'white' framework. This recognition made my students more accountable for their choices. Emphasizing the contingent nature of 'whiteness' and the relational, historical dimensions of *all* racialized identities, I challenged them to investigate and question the unspoken 'white' status quo and to decide whether they were willing to continue choosing to think and act in 'white' ways. Perhaps most importantly, these (de)racialized reading strategies enabled me to expose the destructive components of 'whiteness' without triggering the "retrogressive white identities" (Apple, "Foreword" ix) or the "racist, reactionary thinking" (Gallagher, "Redefining" 33) that occur with growing frequency in our classrooms.

unreading 'race'

In this chapter I have argued that we cannot simply read 'whiteness' into previously unmarked texts and bodies, for to do so recenters 'whiteness,' reinforces the dangerously false belief in separate 'races,' and leads to a 'white' backlash and the development of retrogressive 'white' identities. Instead, reading 'whiteness' can be the first step in a larger process, which enables us, simultaneously, to read and unread 'race' by exposing the arbitrary, politicized nature of all racial identities. It is this twofold process that I have described as (de)racialized reading strategies.

Let me emphasize: I am not in any way recommending that educators engage in a superficial color-blindness that transcends 'race' or ignores the impact of racism. It is, in fact, quite the reverse: (De)racialized readings can be transformational, and expose both the pseudo-universal assumptions we often make about 'whiteness' and the highly destructive implications of racialized thinking and racism. Significantly, however, the reading strategies I propose do so *without* naturalizing or in other ways reifying 'race'—which is, as we all know but too often "forget"—a sociohistorical concept, not a biological fact.

CHAPTER FIVE

Teaching the Other?

"I'm straight; what do *I* know about homosexuality? Besides, I might offend students if I try to include it in my courses."

"But what will people think? What if my Chair finds out? What if parents complain? This is a conservative school . . . ."

These quotations represent two of the major objections many educators give when asked why they don't include writings by or about lesbians, bisexuals, gays, and transpeople[1] in their syllabi. The first obstacle—lack of personal experience—is based on the underlying belief that "Only lesbians [or gays or bisexuals or transpeople] can teach lesbian [or gay or bisexual or trans] issues." This belief, although often motivated by the desire not to appropriate or misrepresent the experiences of others, relies on a simplistic form of identity politics that prevents us from acknowledging the complex interconnections among apparently dissimilar peoples. As Katherine J. Mayberry points out, the "debate over identity-based credibility (teaching what you are versus teaching what you're not) is currently the most visible expression of identity politics in higher education" (3). This debate—which extends far beyond sexuality—revolves around issues concerning personal identity, experience, authenticity, and authority, and leads to a number of interrelated questions: Can we—as educators—teach about our sexual/ethnic/gender/class/national others, or do all attempts to do so inevitably appropriate, stereotype, or in other ways silence these others? To what extent is authority in the classroom dependent on personal identity and experience? Must we "be" gay (or black or Latina or female or 'white,' or so on) to teach material by and about gays (or blacks, Latinas, women, 'whites,' or so on)? This belief that we must "be" what we teach relies on what Frances Smith Foster calls "fraudulent displays of identity politics"[2] and reinforces the already-existing boundaries between what seem to be entirely different peoples.

The second obstacle—fear of public censure—is also understandable, given conservative interest groups' attempts to shape curriculum choices. Yet the only way effectively to challenge the heterosexism and homophobia underlying this fear is to speak out. As Audre Lorde asserts in her well-known essay, "The Transformation of Silence into Language and Action," our silences do not really protect us from the oppressive social systems we fear; they only keep us paralyzed and divided: "It is not difference which immobilizes us, but silence. And there are so many silences to be broken" (*Sister Outsider* 44).

No matter how we identify ourselves, social-justice educators have the responsibility to begin breaking the many silences that paralyze and divide us. To borrow Lorde's words, we must transform these silences first into language, and then action. But in order to do so, we must introduce our students—and at times perhaps ourselves—to a wide range of worldviews and experiences. As Marilyn Frye asserts, "[m]ost of the students and faculty members in U.S. universities in the present era need to be vastly more informed and appreciative of multiplicity, plurality, and diversity, both among and within cultures" (*Willful Virgin* 22). When I first started teaching, I was overwhelmed by the amount of material I felt that I needed to thoroughly learn. Always the over-achiever, I believed that should become an "expert" and know *everything* about all forms of sexuality, ethnicity, gender, class, and all other systems of difference. I soon realized that this type of expertise was impossible, and I modified my approach. My goal now is more modest: I continue learning while recognizing the inevitable limitations in what I know; and I use this expanding knowledge as I re-envision each course I teach. Foster makes a related point in her discussion of literary studies:

> The study of literature, like most areas of intellectual inquiry, has been irrevocably altered by social, political, economic, technical, and perhaps spiritual changes. We will never again be confident that we can recognize or "master" the best thoughts of the best minds. We can neither read nor can our libraries afford to buy *all* the good books available. As teachers and as scholars we need to shift our attention from passing on a tradition to focusing upon ways of knowing, from interpreting the symbols in a text to understanding the entire text as symbolic. (201, her italics)

Foster's assertion can be applied to other academic disciplines as well. Given the proliferation of knowledges in many fields, it is impossible to read, learn, and teach all of the texts that we "should." Like Foster, I believe that this fact does not absolve us of accountability but instead indicates that we must attend more closely to epistemological issues. Hence in *Teaching Transformation* and in my own classrooms, I have focused on relational thinking and reading skills. As I explained in a previous chapter, my pedagogy is shaped by the conviction that education has the potential to give students an increased sense of agency, as well as the skills to become more responsible citizens—capable of effecting individual and collective change. Because I want students to realize that the ideas they're exposed to have material and psychic effects, I structure my syllabi and classroom dynamics to underscore the complex interconnections among personal agency, self-reflection, knowledge, and power. Like Henry Giroux, I believe that "[s]tudents need to learn that the relationship between knowledge and power can be emancipatory, that their histories and experiences matter, and that what they say and do can count as part of a wider struggle to change the world around them" ("Democracy" 25). I want my students to realize that their actions and words matter. They can, if they so choose, positively impact and help to transform existing social conditions.

As in the previous chapters, I will draw on my own teaching experiences to discuss some of the strategies I have found most effective, but in this chapter I will focus primarily on lesbian/gay/bi/heterosexual/trans material. These strategies—which include framing each course, fostering intellectual agency, (non)naming tactics, focusing on multiple linked issues, questioning conventional definitions, and historicizing the categories—do not apply specifically and exclusively to sexuality–gender-related identities and issues. However, I have decided to foreground sexuality for several reasons: First, I developed these teaching tactics as I tried to incorporate sexuality into my introductory-level courses filled with conservative, narrow-thinking students. Second, I want to challenge the commonly held though rarely examined assumption that ethnicity, 'race,' and (sometimes) gender constitute the only components of multicultural theory and praxis. Moreover, because sexuality is often viewed as a highly private, taboo topic that should not be discussed in public, and certainly not in academic settings, it can be a very difficult pedagogical terrain in which to maneuver—no matter how we self-identify.

strategy #1: framing each course

Framing each course—how we present ourselves and our material—begins with the syllabus and includes course objectives, teaching methods, and self-definition. Although it seems almost too obvious to state, I find that my teaching is much more effective when I have clearly delineated my course goals both to myself *and to my students*. As Mayberry notes, "all responsible arguments must work from a clear identification of what one hopes to achieve through teaching and what methods are most likely to effect these goals. These do not sound like revolutionary questions, yet too few of us—trained as disciplinary scholars rather than educators—have seriously entertained them" (6). The need for clearly stated objectives and goals is especially important when including potentially controversial material.

I learned this lesson the hard way, when Mary,[3] a graduate teaching assistant, was accused of assigning inappropriate material in a composition course. The essay in question was a short personal narrative written by a person who identified as bisexual and Native American. The piece itself was an acceptable text for the course; however, one student accused Mary (an outspoken activist) of politicizing the classroom. With the assistance of several conservative community groups, this angry student created a campus-wide uproar and worked vigorously to dismantle our university's recently developed Diversity Plan. When Mary came to talk with me about the situation, I asked her why she had selected this particular piece. I prefaced my question by assuring her that, for a number of reasons, I found the essay to be totally acceptable for use in composition studies. (Its personal narrative format, its tone, its challenge to monolithic identities, and its structure made it extremely appropriate for introductory-level students. However, I did not explain *why* I found it acceptable because I did not want to influence her reply.) Although Mary probably knew these reasons at an almost unconscious level, she could not clearly articulate them to me. Instead, she made vague claims about the essay's usefulness. Not surprisingly, then, when she assigned the essay to her class she did not explain how it fit with the course goals. Had she done so, she would have made it more difficult for the student to object. Already vulnerable because of her youth, her gender, her sexuality, her politics, and her status as a graduate teaching assistant, Mary made herself an open target! This experience taught me two lessons: (1) it is crucial to reflect carefully about text selection and course goals; and (2) it is equally important to share these reflections with my students both on my syllabi and during class discussions.

Providing students with a framework for readings and discussions serves two purposes. First, it allows me to anticipate possible objections before they arise. Both in my syllabi and in my introductory lectures I inform students that my pedagogy is shaped by a transformative-multicultural perspective. Since my students' only knowledge of multiculturalism is derived from the media (Cinco de Mayo festivals, Nikelodeon TV, and other such "celebrations" of diversity), I carefully distinguish between melting-pot multiculturalism's benign pluralism, separatist multicultural-ism's divisiveness, and transformational multiculturalism's sometimes uncomfortable exploration of relational differences. Whereas the first two types of multiculturalism incorporate diverse perspectives without inter-rogating the status quo—the heterosexual, eurocentric, masculinist norm—transformational multiculturalism examines this norm; it explores both the many power struggles informing U.S. culture and the ways sexuality, gender, culture, ethnicity, and other variables are shaped by sociohistorical forces and representational strategies.[4] By analyzing a broad range of interlocking worldviews, issues, and experiences, I explain, we will investigate how mainstream definitions create hierarchical social systems that encourage everyone—from the notorious 'white' middle-class (heterosexual) male to the sexual/gendered/class/ethnic other—to think and act in certain ways.

Second, providing students with a framework for class discussions and readings enables me to acknowledge their agency and make them more responsible for their own education. They can evaluate course objectives, perspectives, and anticipated outcomes. And if, as I believe, language is or can be performative, then the act of *telling* students what I hope they will learn from a specific course or text makes it more likely that they will be open to the particular lessons, perspectives, and skills I envision.[5] I explain to my students that by requiring them to engage with multiple, interconnected perspectives I hope to encourage their relational thinking skills—their ability to employ analysis, imagination, and self-reflection in conjunction. In the classroom, relational thinking will enable them to recognize and investigate how the texts, theories, and knowledge systems we explore are shaped by personal, cultural, and historical issues. I explain that because language is never neutral, the words we encounter are always context-specific and thus come to us already infused with meaning and value. This awareness of the inevitably intersubjective nature of language encourages students to investigate how writers have appropriated and shaped the words they use to fulfill their own intentions, intentions that are never purely their own. They will learn to analyze each text's underlying assumptions, use their own knowledge and experience to evaluate these worldviews, and draw connections among the various texts and worldviews we explore. I also inform students that relational thinking can extend beyond the classroom to provide them with new perspectives about many issues, ranging from the advertisements they encounter everyday to the stereotypes circulating in contemporary culture.

Recognizing that labels like *'black'/'white,' masculine/feminine,* and *self/other* are themselves mutually constituted and thus interconnected, they will learn to challenge restrictive identity categories and recognize the connections among themselves and differently situated individuals and groups.

I also insist on the vital role holistic-critical thinking plays in classroom assignments and discussions. In my first years of teaching, I emphasized critical thinking but did not include the word *holistic* in my definition. I found that although *critical thinking* and *criticism* are not synonymous, my students often conflated the two. Because they assumed that critical thinking was entirely negative, oppositional, and divisive, they approached every assigned text in a binary-oppositional, closed fashion: Adopting an attack mode of sorts, they focused *only* on limitations and weaknesses. By using the term *holistic-critical thinking,* I invite students to take a more relational, both/and approach. As I define the term, holistic-critical thinking is premised on interconnectivity, coupled with confidence in each person's intellectual[6] potential. Holistic-critical thinking combines independent thought with a relational perspective. Encouraging a both/and approach, holistic-critical thinking enables us to explore how apparently isolated peoples, situations, ideas, and events function as parts of a larger whole.[7]

strategy #2: fostering intellectual agency

I want students to know that I greatly value their intellectual agency. Whether teaching undergraduate or graduate courses in composition, literature, or women's studies, I encourage them to become active learners. I explain that if they simply accept what others tell them they become victims—controlled by others' ideas, dominated by others' laws. Throughout the semester I encourage students to think for themselves and put their experiences into dialogue—both with the texts we read and with other students' perspectives. In addition to using discussion questions,[8] I include statements such as the following in each syllabus:

I encourage you to use personal opinions and references to your own experiences during our class discussions; however, you must be sure to relate your opinions and experiences to the assigned texts. In other words, I'm interested in how you use the texts to support your opinions, and how you use your personal experiences and opinions to challenge and support the texts.

By thus emphasizing the reciprocal interaction between their own experiences and the course texts and perspectives, I invite students to recognize and explore the many interconnections between the words they read and the world in which they live. (This recognition is the first step to enacting change.)

I assure students that I do not penalize them for their views, and I try to create an environment that values differences of opinion. My syllabi contain statements such as the following:

> Because it's important that we all respect each other's needs, values, beliefs, and views, our discussions will be guided by open-mindedness and respect for each person's opinions, no matter how different they are from our own. Or as June Jordan states, "Agreement is not the point. Mutual respect that can accommodate genuine disagreement is the civilized point of intellectual exchange, particularly in a political context."

> Understanding the assigned texts does not mean that you have to adopt all or any of the viewpoints represented; however, it does require that you read, think about, and discuss the material.

These statements serve two interrelated purposes: First, they emphasize the importance of open-mindedness and acknowledge the fact that we have a variety of opinions and views. By so doing, they create a space in class discussions for students to voice dissenting views and thus demonstrate that I am not simply proselytizing: I do not want my students to adopt some type of womanist/feminist party line! Second, these statements allow me to insist on reciprocal respect during class discussions. Just as I respect my students' opinions, beliefs, and perspectives, I want them to respect those of their classmates.

To further emphasize the importance of intellectual agency, I openly question conventional education's hierarchical power structures. I explain that I will not adopt the role of the "expert" who tells students what to think about the course material. Instead, by troubling the conventional understandings of terms such as *classic, literary value, homosexual, heterosexual, masculine, feminine, feminist, womanist, race, man,* and *woman,* I hope to encourage students' own holistic-critical thinking skills. Rather than unthinkingly accept the dominant culture's mandates and inscriptions, I want them to begin (re)defining these terms for themselves.

strategy #3: tactical (non)naming

At times I even invite students to question my own sexuality and ethnicity/ 'race' by engaging in what I call *tactical (non)naming*. That is, I make students aware of the fact that I do *not* identify my sexual or racial affiliations, and I inform them of the complex rationale behind my decision. I carefully tailor my explanation, based on the specific course topic, student body, and other factors. One semester, for example, I told my introductory women's studies class that I had consciously chosen not to discuss my own sexual identity, and I explained that I hoped to avoid what I see as two possible dangers in teaching queer material: On the one hand, if I identified as "straight," it would be difficult not to establish an us/them polarity where "we" heterosexuals examine "those" sexual other(s); on the other hand, if I identified as "gay," "lesbian," "bisexual," "homosexual," or "queer,"[9] I might be expected—or tempted—to speak for all gays, lesbians, bisexuals, homosexuals, and/or queers—implying that I could give students an authentic insider's view that ignores the enormous diversity within these groups. I discussed other issues related to incorporating sexuality into the course material, including the importance of lesbian-identified role models, the politics of coming out and the academic community's homophobia, my own fears that I'd be labeled as lesbian (whether or not I actually was), and the temptation to refer to a fictionalized or real-life boyfriend, male lover/partner, or husband.[10] This staged (non)confession led to a lively discussion of homophobia and other internalized and external forms of oppression.

I am not advocating that instructors should never disclose their sexuality in the classroom. As the term *tactical (non)naming* implies, the decision to remain silent concerning my own sexual identity was a carefully chosen tactic based on my assessment of the particular students who signed up for my introductory women's studies course that particular semester. Many of them were rather conservative education majors, and I wanted to encourage them to incorporate lesbian/gay/bi material into their teaching in non-homophobic ways. Thus I prefaced my open refusal to self-identify with a discussion of classroom-based homophobia. All too often I have seen heterosexually identified educators or speakers (and others passing as straight) nervously preface their remarks on homosexuality with comments designed to assure their listeners that *they* are *not gay*. These disclaimers strike me as highly destructive, and I wanted my per- formance to serve as an alternate model. However, as I explain to my students, my point is not that instructors should *never* identify their sexuality but rather that different contexts require different approaches; tactical (non)naming is one possibility among others.

I also use this tactical (non)naming when discussing texts and issues related to ethnicity, cultural identities, or the arbitrary nature of U.S. racialized categories and divisions. For instance, when I teach surveys of U.S. literature I preface discussions of slave narratives with a brief summary of historical and contemporary issues related to miscegenation, passing, and the highly racist political and economic motivations informing the development of seventeenth-, eighteenth-, and nineteenth- century racialized categories. I then challenge students to guess my ethnicity/'race.' If they fail to do so, I briefly discuss my own family history, the rule of hypodescent that marks me as "black," my family's various attempts to celebrate, ignore, and deny this racialized designation, and the fact that many people who seem to be 'black,' 'white,' or any other single 'race' are actually mixed.[11] My point, I explain, is that appearances can be deceptive and, quite possibly, not one of us is "unmixed." Indeed, the implicit belief in entirely distinct 'races' implies a false sense of racial purity, for we could all be described as multiracial. Furthermore, the suggestion that we can automatically identify ourselves and others according to 'race' is inaccurate and misleading. Since appearances are so often misleading, this tactic could be adopted by educators of many colors.[12]

strategy #4: focusing on multiple interlinked issues

One of the most useful strategies I employ involves exploring diverse sets of issues simultaneously. By discussing sexualities in conjunction with ability/health, class, ethnicity/'race,' gender, region, and/or religion, I normalize this identity category: sexuality becomes one aspect of the complex systems of difference that shape social actors, texts, and traditions in contemporary U.S. cultures. This emphasis on multiple systems of difference involves three interrelated techniques.

First, I avoid focusing exclusively on homosexuality and instead structure my courses to examine the similarities and differences among multiple forms of sexuality. At times, I juxtapose "straight," "queer," and "trans" texts in order to stimulate comparative analyses; I have taught Mary Helen Ponce's *The Wedding* in conjunction with Rita Mae Brown's *Rubyfruit Jungle* and Joanna Russ's *The Female Man*; E. Lynn Harris's *Invisible Life* with Ralph Ellison's *Invisible Man*; Paule Marshall's *Praisesong for the Widow* with Audre Lorde's *Zami: A New Spelling of*

My Name; and Dorothy West's *The Living Is Easy* with Terri de la Peña's *Margins*. I like to include works—such as those by Ana Castillo, E. Lynn Harris, Randall Kenan, Cherry Muhanji, April Sinclair, Alice Walker, and Walt Whitman—that present a variety of sexualities and genders. I have found Denise Chávez's *Last of the Menu Girls*, Toni Morrison's *Sula*, Sarah Orne Jewett's *Country of the Pointed Firs*, Jackie Kay's *Trumpet*, and Marge Piercy's *Woman on the Edge of Time* to be especially effective in challenging students' rather simplistic definitions of lesbianism, heterosexuality, and female-gendered identity issues; these texts' ambiguous depictions of women's sexualities, coupled with their portrayal of autonomous female characters and unconventional male characters, have provoked fascinating classroom discussions. By thus exploring same-sex and male–female relationships simultaneously, neither homosexuality nor heterosexuality can be privileged. (It is, in fact, almost the reverse: As I will explain later, by examining heterosexuality as one sexual preference among others, I can de-naturalize it.)

Second, I rarely discuss sexualities in isolation from other systems of difference. As mentioned previously, I structure my syllabi in ways that enable us to examine lesbian, gay, bisexual, and heterosexual identities in the context of other issues, such as definitions and stereotypes, the importance of self-naming, and the various forms of oppression (internalized, horizontal, and vertical) often experienced by members of so-called "minority" racial groups and women of all colors. For instance, in a graduate-level seminar on 'Race,' Gender, and Literature I integrated issues related to sexuality and writing into our examination of texts from a wide variety of ethnic/cultural perspectives. We explored both the lesbian subtexts and the critiques of heterosexual relationships in writings by Sui Sin Far, Sarah Orne Jewett, Alice Dunbar-Nelson, and Mary Wilkins Freeman;[13] analyzed the ways self-identified lesbian/gay writers such as Gloria Anzaldúa and Judy Grahn create characters who can pass as straight; and applied Eve Sedgwick's theory of homosocial desire to works by Henry James, Rolando Hinojoso, Scott Momaday, and William Faulkner. Similarly, when I teach Whitman, James, Dickinson, or other nineteenth-century canonical writers assumed by some scholars to be homosexual, I discuss the scholarship that explores these writers' sexuality and ask students in what ways—if any—sexual preferences and identities might influence their works.[14] When I teach Introduction to Women's Studies I do not include a unit or readings that focus exclusively on sexuality (or any other identity category). Instead, we discuss a variety of identity issues in the context of thematic concerns. So, for example, while discussing social movements I include readings, discussions, and films that explore the ways peoples' sexual identities, gender expressions, and self-perceptions have been shaped by feminist movement;[15] and in a unit on childhood, I include readings about transchildren.[16] When I was asked to teach a graduate-level women's studies course on queer theory, I designed a course called *Transgressive Identities* that explores both queer theory and critical 'race' theory. Rather than focus on sexuality or 'race,' we investigated these categories while looking at additional issues, like the relationship between language, personal/collective identities, and material reality; the impact storytelling and other forms of language have on material reality; and the roles storytelling can play in changing individual and collective consciousness and in other ways bringing about progressive social change. (See appendix 6.)

Third, and closely related, I create dialogic groupings of carefully chosen texts. That is, I select writings that explore complex, interrelated issues from a wide range of literary, cultural, and sexual perspectives; and I encourage students to read these texts in conversation with each other. I structure syllabi, assignments, discussions, and lectures in ways that enhance dialogue, requiring students to grapple with conflicting yet overlapping worldviews; and I organize courses thematically, focusing on common issues that we explore from a variety of perspectives. I frequently use writings by Anzaldúa, Cherríe Moraga, Becky Birtha, Lorde, and other self-identified U.S. queers and lesbians of colors because their experiences so effectively illustrate the complexity of intersectionality; as Moraga explains, their "very presence violates the ranking and abstraction of oppressions." Similarly, the stories, essays, narratives, and poems collected in anthologies such as *This Bridge Called My Back: Writings by Radical Women of Color*; *Making Face, Making Soul/Haciendo Caras: Creative and Critical Perspectives by Women of Color*; *Brother to Brother: New Writings by Black Gay Men*; and *this bridge we call home: radical visions for transformation* do not separate gender from sexuality, ethnicity, or class. These works provide students with concrete examples of the interlocking systems of oppression that marginalize people who—because of their sexuality, gender, ethnicity, and/or economic status—do not belong to the dominant cultural group.

This multi-perspective approach shapes students' perceptions in significant ways. By negotiating among commonalities and differences, and exploring broad issues that reflect the voices of numerous groups, these transcultural dialogues enable us to examine issues crossing gender, sexual, ethnic, and class lines *without* reinforcing stereotypes concerning the various identity categories we encounter. Introducing students to a wide range of worldviews—including the experiences of self-identified

African-American, Arab American, Asian American, Chicana, Native American, and 'white' lesbians, bisexuals, heterosexuals, transpeople, and queers—effectively challenges their preconceptions about their sexual, ethnic, gender, or class other(s). It becomes difficult, if not impossible, for students to accept monolithic definitions. Compelled to reexamine their own preconceptions, as well as social representations of these various categories of difference, they recognize that just as there's no unitary, all-encompassing description of "woman" or "Chicana," there's no single definition of "lesbian," "bisexual," "gay," or "trans."

Most importantly, the negotiations among commonalities and differences, coupled with my emphasis on a multiplicity of differences, prevents students from equating difference with deviation. An exclusive focus on any single category—be it sexual preference, gender, culture, ethnicity, or class—inadvertently reifies the unexamined (Euro-American, male, heterosexual) norm and, by extension, implies the abnormality of the (non-Euro, nonmale, nonheterosexual) group under examination. As Paula Rothenberg asserts, "[w]here white, male, middle-class, European, heterosexuality provides the standard of and the criteria for rationality and morality, difference is always perceived as deviant and deficient" (43). Discussing same-sex relationships in the context of all forms of sexuality makes it difficult, if not impossible, for students to dismiss the former as "abnormal." And by examining diverse forms of sexuality in conjunction with other systems of difference, sexual identities can be seen as one aspect of the complex social identity categories that construct human beings and shape literary texts in contemporary U.S. cultures. I find that this approach can work with even some of my most conservative students.

strategy #5: questioning conventional definitions

Frequently, students' automatic acceptance of heterosexuality as normal, as well as their conceptions of lesbian, gay, bisexual, and queer identities as deviant, rely on status-quo stories based on false information and homophobic stereotypes: "Heterosexuality is natural;" "All lesbians hate men;" "All gay men are effeminate;" and so on. Like the status-quo stories I described in chapter one, these beliefs prevent students from recognizing commonalities between themselves and those they view as "other." To challenge these inaccurate assumptions, I de-naturalize all forms of sexuality—including the heterosexual norm—by discussing the ways sexualities are constructed in U.S. culture and asking students to consider how these constructions impact our explorations. The issues we explore vary, based on the particular course and student population. Some of the issues we have examined include the following:

- How have the media and religious institutions portrayed homosexuals, heterosexuals, and sexuality itself?
- What are the consequences of these images? How might they impact self-identified lesbian/gay/queer people?
- Why do you suppose that ethnic-specific canons often excluded or downplayed non–heterosexually identified texts and writers?[17]
- Why might many people find it difficult to discuss same-sex desires in literary texts?
- Why has HIV/AIDS been defined as a gay disease? How might this media-constructed image adversely affect AIDS research and treatment?
- Given some other cultures' more lenient attitudes toward same-sex relationships, why might contemporary western societies see homosexuality as so threatening?[18]
- How does homosexuality jeopardize the status quo, and how does heterosexuality reinforce it? How does heterosexual privilege reward those who conform to the dominant culture's proscriptions?
- How does homophobia lock all people into rigid gender roles?
- How do accusations of lesbianism prevent women from forming bonds with each other, or from identifying as feminists?[19]

These questions are, of course, context-specific. The particular issues I explore in each class are shaped by course content, previous class discussions, and levels of student resistance.

Like status-quo stories about 'race,' status-quo stories about sexuality and gender posit fixed, unchanging, natural identities. Often students assume that homosexuality and heterosexuality reflect innate character traits: Each person is born with a single, permanent sexual identity. Although some people would agree with this view, I stress that no one knows for sure how sexual identities develop, and often sexuality changes within the course of a person's life. Students are surprised to learn that many theorists see heterosexuality, as well as homosexuality, as a matter of personal choice. Marilyn Frye, for example, challenges self-identified heterosexual feminists to reflect on their own sexual preference, to be "as actively curious about how and why and when they became heterosexual as [she has] been about how and why and when [she] became Lesbian [sic]" (*Willful Virgin* 55).

John Stoltenberg offers a related challenge in his discussion of gendered categories of sexuality. He maintains that "we are not born belonging to one or the other of two sexes. We are born into a physiological continuum on which there is no discrete and definite point that you can call 'male' and no discrete and definite point that you can call 'female'" (28). According to Stoltenberg, it is culture—not nature—that has created the absolute division of all human beings into two categories. He argues that this division leads to restrictive notions of sexuality, where those people (with penises) identified as "male" are trained to experience physical pleasure and, by extension, heterosexual male sexuality in restrictive ways:

> So-called male sexuality is a learned connection between specific physical sensations and the idea of a male sexual identity. To achieve male sexual identity requires that an individual *identify with* the class of males—that is, accept as one's own the values and interests of the class. A fully realized male sexual identity also requires *nonidentification with* that which is perceived to be nonmale, or female. (33–34, his emphasis)

I use writings by Max Wolf Valerio, Riki Wilchins, Diana Courvant, Jody Norton, Jamison Green, Henry Rubin, and Kate Bornstein to challenge and expand students' understanding of trans identities and issues. In addition to exploring the similarities and differences between "FTM" and "MTF" transpeople, these authors demonstrate a wide variety of perspectives. Whereas some view trans identities as innate and cross-historical, others view them as entirely constructed. According to Jody Norton, for instance, transchildren can be found throughout history: "there appear always to have existed some children who have either been unable or unwilling to be 'boys' or 'girls' in the conventional ways their bodies suggested they should" (148). Riki Wilchins offers a very different perspective, asserting:

> While I recognize how important it is to produce histories and sociologies of transpeople, I am wary of anything that might cement the category more firmly in place. I'd also like to investigate the means by which categories like *transgender* are produced, maintained, and inflicted on people like me. It's not so much that there have always been transgendered people; it's that there have always been cultures which imposed regimes of gender. It is only within a system of gender oppression that transgender exists in the first place. (67, her emphasis)

My goal with this fifth strategy is, in part, to challenge students' tendency to make snap judgments about these complex issues.

Essays, stories, and poems by Allen, Moraga, Anzaldúa, Lorde, Adrienne Rich, Birtha, and other self-identified lesbians further challenge students' assumptions by illustrating the diverse and sometimes contradictory ways lesbians reconstruct their own sexual identities. For example, Anzaldúa claims that she "made the choice to be gay" and alludes to a time in her life when she was heterosexual (*Borderlands/La Frontera* 19), whereas Moraga maintains that she has always been queer (*Loving in the War Years* 52). Surprisingly (especially for students), Moraga's conviction that sexual identities are innate leads her to insist that Anzaldúa's assertion does not indicate a shift from heterosexuality to lesbianism. According to Moraga, Anzaldúa was (always) already lesbian; her "choice to be gay" refers only to her political decision to identify openly as gay ("Algo secretamente amado"). Yet Anzaldúa disagrees with Moraga's interpretation and insists that she views identity—including sexuality—as quite malleable and open to change in the course of one's life.[20] By discussing this disagreement with students, I invite them to explore their status-quo stories about lesbians and sexual identity categories more generally.

I challenge students' conceptions of sexuality and gender even further by mentioning some of the unresolved (and probably unresolvable) theoretical issues queer theory explores: What does it mean to be lesbian, bisexual, gay, or queer? Is it simply a matter of sexual object choice? If so, how do we account for self-identified lesbians and gay men who occasionally have sex with people from the opposite gender?[21] Say, for example, a self-identified gay man has sex with a self-identified lesbian woman: have they become (temporarily) heterosexual? Are they even performing a heterosexual act? Do the terms *lesbian* and *gay* refer to specific lifestyles? Is there, for example, a "gay sensibility"? a "lesbian" worldview? a "lesbian ethics"?[22] What is the relationship between queer theory and trans issues?

I want my students to realize that there are no fixed definitions of lesbian, gay, bisexual, trans, or queer identity. Indeed, the word *lesbian* is itself extremely problematic. Often, people automatically believe that the term indicates a unitary, essentialized core identity, yet this interpretation is too restrictive. As Ruth Ginzberg implies, it overlooks specific "*acts, moments, relationships, encounters, attractions, perspectives, insights, outlooks, connections, and feelings*" that could also be considered "lesbian" (82, her emphasis). Not surprisingly, then, some self-identified lesbian theorists such as Sarah Hoagland and Marilyn Frye refuse to define the term, explaining that lesbians don't exist in phallocentric conceptual systems. By thus locating lesbians *outside* patriarchal structures, they can claim that lesbians represent a new non-masculine subjectivity. Monique Wittig makes a similar point in "The Straight Mind" when she distinguishes between *woman* and *lesbian*. She asserts that " 'woman' has meaning only in heterosexual systems of thought and heterosexual economic systems. Lesbians are not women" (32). I further underscore the diverse—and often conflicting—ways lesbians have defined themselves by discussing the positive and negative implications of the *woman-identified woman* and Adrienne Rich's *lesbian continuum*, as well as the debates between nonlesbian feminists, feminist-lesbians, and other lesbians.[23]

Many self-identified lesbians of colors take this interrogation of lesbian identity even further by exposing and rejecting its eurocentric bias. Anzaldúa, for example, views the words *lesbian* and *homosexual* as 'white,' middle-class labels and often adopts expressions like *tejana tortillera, mita' y mita,' patlache*, and *mestiza queer* to describe herself.[24] As she repeatedly emphasizes the cultural- and class-specific dimensions of her sexual identity, she destabilizes ethnocentric (Anglo, middle-class) definitions of homosexuality and (re)defines herself and other dykes as "forerunner[s] of a new race, / half and half—both woman and man, neither / a new gender" (*Borderlands/La Frontera* 194). And in a 1990 interview Paula Gunn Allen draws on her Native American worldview to redefine lesbianism as a form of trickster identity, which she associates with transformation. For Allen, lesbians and gay men exemplify "the sacred moment, the process of changing from one condition to another—*life-long liminality*" (qtd. in Caputi 56, Allen's emphasis).

strategy #6: historicizing the categories

I further challenge students' conceptions of lesbian, bisexual, trans, and gay identities by exposing the ahistorical nature of their status-quo stories about sexuality. Like their ahistorical views of 'race,' students believe sexuality to be permanent and unchanging. Generally, they assume that contemporary definitions of sexuality and sexual identity—especially the rigid hetero/homo binary—are unchanging categories that have existed in every historical period. Yet these terms, as well as modern western culture's interpretations of sexuality, are fairly recent inventions. As David Halperin asserts, "homosexuality, heterosexuality, and even sexuality itself [are] relatively recent and highly culture-specific forms of erotic life—not the basic building-blocks of sexual identity for all human beings in all times and places, but peculiar and indeed exceptional ways of conceptualizing as well as *experiencing* sexual desire" (9, his emphasis).

Significantly, students seldom realize that the words *homosexuality* and *heterosexuality*—as well as the practice of classifying specific types of people according to their sexual preferences—did not exist before the 1890s. But as Arnold Davidson persuasively argues, until the second half of the nineteenth century sexuality was defined exclusively on the basis of anatomical sex; it was not until the "emergence of the psychiatric style of reasoning" that the idea of sexual identities, conceptualized as "a matter of tastes, aptitudes, satisfactions, and psychic traits," developed. It is not that people were not engaging in homosexual acts before this time but rather that these practices or behaviors did not indicate specific personality types or psychosexual identities. For example, sodomy—which had been perceived as a specific sex act that could involve people of any gender—was transformed into a particular type of person with distinct psychological traits: the sodomite or homosexual (18–21).[25]

speak out!

I opened this chapter by mentioning two obstacles self-identified heterosexuals and others often confront as they consider incorporating lesbian/gay/bi/trans issues into their syllabi: (1) lack of information, and (2) fear of public censure. I have devoted most of this chapter to a discussion of the former, and to some extent this focus has been intentional. After all, it's easier to compensate for one's own lack of knowledge than it is to deal with other people's homophobia and heterosexism. However, the two issues are intimately related. As Lorde suggests in "The Transformation of Silence into Language and Action," both ignorance and fear have their source in the dynamics of difference. More specifically, our misperceptions of those whom we view as "different" often prevent us from establishing bonds with our sexual, ethnic/'race,' gender, bodied, and/or class others. To be sure, speaking out does not automatically make our fears go away. It does, however, enable us to begin redefining categories and forming political alliances with each other. Thus Lorde challenges her audience to overcome

the mockeries of separations that have been imposed upon us and which so often we accept as our own. For instance, "I can't possibly teach Black women's writing—their experience is so different from mine." Yet how many years have you spent teaching Plato and Shakespeare and Proust? Or another, "She's a white woman and what could she possibly have to say to me?" Or, "She's a lesbian, what would my husband say, or my chairman?" . . . And all the other endless ways in which we rob ourselves of ourselves and each other. (*Sister Outsider* 43–44)

Like Lorde, I am firmly convinced that we must challenge the oppressive social systems and the culturally constructed boundaries that prevent us from recognizing our complex interconnections with apparently dissimilar women and men. But in order to do so, we must risk speaking out—despite our fears. By so doing, we begin bridging the differences that seem to divide us.

Transformational multiculturalism can play an effective role in this process, for it compels educators and students to leave the safety of preestablished identity concepts and recognize the limitations of all cultural inscriptions. Exploring texts written from sexual, ethnic, gender, and/or class perspectives that differ from our own provides us with opportunities to discover new points of commonality among ourselves and others. Instead of limiting us to monologic conversations on single-issue topics such as homophobia, sexism, racism, or other forms of oppression, transformational multiculturalism enables us to examine issues crossing gender, ethnic, sexual, and class lines. Exploring broad issues capable of reflecting the voices of numerous groups, transformational multiculturalism demonstrates that it is the refusal to acknowledge and accept differences—rather than the reverse—that erects rigid boundaries between identity categories and thus prevents open dialogue among disparate readers. Although we must acknowledge and help students explore the borders—the representational strategies, historical and contemporary sociopolitical contexts, and belief systems—that make each group distinct, we should not present these boundary lines as impenetrable barriers separating people of diverse backgrounds. It is, in fact, the reverse: As we examine these boundaries we discover and invent new commonalities that neither erase nor solidify the differences between self and other.

CONCLUSION

May We Dream New Worlds into Being: Transforming Status-Quo Stories

[T]o a very great extent we dream our worlds into being. For better or worse, our customs and laws, our culture and society are sustained by the myths we embrace, the stories we recirculate to explain what we behold. I believe that racism's hardy persistence and immense adaptability are sustained by a habit of human imagination, defective rhetoric, and hidden license. I believe no less that an optimistic course could be charted, if only we could imagine it.

Patricia Williams[1]

Like Patricia Williams, I believe that we play a crucial role in shaping the worlds we inhabit. All too often, however, we do not recognize this creative power. Born into a reality filled with customs, stories, and myths that have already been recirculated countless times, we're taught in ways both subtle and overt that these status-quo stories are permanent, unchanging reflections of "The Truth." As I explained in chapter one, status-quo stories reaffirm and reinforce the existing social system, making reality monolithic and fixed.

Status-quo stories are divisive. Teaching us to break the world into parts and label each piece, they contribute to the "defective rhetoric" Williams describes in my epigraph—a language and belief system that facilitate not only racism but also sexism, homophobia, and other forms of oppression. These labels and their defective rhetoric, in turn, shape the ways we view ourselves and each other; by so doing, they further naturalize the status-quo stories they support. Status-quo stories foster dualistic worldviews, self-enclosed individualism, and other forms of binary-oppositional thinking.

Not surprisingly, then, status-quo stories stunt our imaginations and prevent us from envisioning alternate possibilities—different ways of living, acting, and arranging our lives. Status-quo stories deny personal agency, complicity, and accountability. When we live our lives according to status-quo stories, we're convinced that the way things are is the way they always have been and the way they must be. This belief becomes self-fulfilling: We do not try to make change because we believe that change is impossible to make. In this way, status-quo stories naturalize racism and other interlocking forms of social injustice. Because we believe that we are powerless to change either ourselves or our worlds, we recreate the status quo as we recirculate these outworn, destructive stories. With each retelling, the stories become more deeply embedded, more solid and fixed.

I, too, have been seduced into believing that status-quo stories accurately represent reality. Although they created a world I intensely disliked, I believed myself powerless to change myself or this world. It was not until reading Ralph Waldo Emerson, Audre Lorde, Paula Gunn Allen, and Gloria Anzaldúa that I learned to "see through"[2] the status-quo stories that held me in thrall. These authors unlocked my imagination and, through their own bold writing, outrageous imaginations, and willingness to take risks, invited me to dream new worlds into being, to create new stories about my worlds.

As this book demonstrates, I'm working to invent a story of multicultural-feminist/womanist pedagogy and relational reading, or what I call *transformational multiculturalism*. I hope to replace status-quo stories' fragmented thinking and oppositional worldviews with a holistic, relational worldview that insists on our interconnectedness. Transformational multiculturalism is holistic; it encourages me to develop nonbinary-oppositional epistemological and pedagogical methods that work in the service of social justice. Positing interconnectivity, I work to reconfigure dualistic relationships (between "self"/"other," "us"/ "them," "oppressor"/"oppressed," and so on) into nonbinary forms.

In my teaching and, indeed, throughout my life, transformational multiculturalism enables me to insist on personal and collective agency. We are all implicated, to varying degrees and in various power-inflected ways, in the status-quo stories currently shaping our lives. This complicity makes us accountable; we must work to investigate and transform the 'white'-supremacist masculinist framework that sustains status-quo stories' power. I maintain the possibility of enacting this change—on multiple interrelated levels, beginning with myself and moving outward to encompass readers, students, and the larger world. But I'm not overly naive! Although change is possible, it is neither automatic nor guaranteed. It requires hard work, painful self-reflection, and tremendous optimism.

My relational story of transformational multiculturalism is still in process and open to further change. As I retell and revise this story, I hope to establish new transcultural dialogues, multidirectional conversations in which all individuals and groups can be altered through our encounters. In classrooms, these transcultural convers(at)ions represent dynamic, synergistic negotiations among differently situated peoples, traditions, categories, and/or worldviews. These negotiations occur on multiple levels, ranging from the texts we read to our embodied experiences as students and teachers. When I attempt to enact transcultural dialogues, I do not ignore the many differences among us but instead use these differences to generate complex commonalities. Through this process of translation and movement across a variety of social categories—including but not limited to those established by ability/health, class, ethnicity, gender, 'race,' region, religion, and sexuality—I invite my students and readers to engage in a series of convers(at)ions that destabilize the rigid boundaries between apparently separate individuals, traditions, and cultural groups.

With these transcultural dialogues, I attempt to chart a more "optimistic course."

APPENDIX 1

Dialogue: Some of My Presuppositions

Here are some of the presuppositions for our class discussions:

1. *Social injustice exists.* People are not treated equitably. We live in an unjust society and an unfair world. Although "liberty," "equality," and "democracy" are radical ideas with great promise, they have not yet been fulfilled. Oppression (racism, classism, sexism, etc.) exists on multiple seen and unseen levels.

2. *Our educations have been biased.* The eurocentric educational systems, media outlets, and other institutions omit and distort information about our own groups and those of others. These hidden mechanisms sustain oppression, including an often invisible and normative 'white' supremacy. Not surprisingly, we all have "blank spots," desconocimientos (Anzaldúa), and so forth.

3. *Blame is not useful, but accountability is.* It is nonproductive to blame ourselves and/or others for the misinformation we have learned in the past or for ways we have benefited and continue benefiting from these unjust social systems. However, once we have been exposed to more accurate information, we are accountable, and we should work to do something with this information—perhaps by working toward a more just future.

4. *"We are related to all that lives."** We are interconnected and interdependent in multiple ways, including economically, ecologically, linguistically, socially, spiritually.

---

*Inés Hernández-Ávila. "An Open Letter to Chicanas: On the Power and Politics of Origin." *Reinventing the Enemy's Language: Contemporary Native Women's Writing of North America.* Ed. Joy Harjo and Gloria Bird. New York: W.W. Norton, 1997. 237–46.

5. *Categories and labels shape our perception.* Categories and labels, although often necessary and sometimes useful, can prevent us from recognizing our interconnectedness with others. Categories can (a) distort our perceptions; (b) create arbitrary divisions among us; (c) support an oppositional "us–against–them" mentality that prevents us from recognizing potential commonalities; and (d) reinforce the unjust status quo. Relatedly, identity categories based on inflexible labels establish and police boundaries—boundaries that shut us in with those we've deemed "like" "us" and boundaries that shut us out from those whom we assume to be different.

6. *People have a basic goodness.* People (both those we study and class members) generally endeavor to do the best they can. We will all make mistakes, despite our best intentions. The point is to learn from our errors. In order to learn from our errors, we must be willing to listen and to speak (preferably, in this order!).

Listening with Raw Openness

Listening is a crucial yet too often overlooked element in effective class discussions and other forms of dialogue. Below are some suggestions that, if we all practice, will enhance class discussions. I describe this process as deep listening, or "listening with raw openness."

1. *Deep listening entails respect for each speaker's "complex personhood"* (Cervenak et al.). As we listen, we remind ourselves that each individual we encounter has a specific, highly intricate history, an upbringing and life experiences that we cannot fully know. We don't know the forces that shaped her and, at best, we can only partially ascertain her intentions and desires. Our understanding is always partial and incomplete.

2. *Deep listening entails the willingness to be vulnerable: opening to others' perspectives, acknowledging the possibility of error, and willing to change.* As Paula Gunn Allen suggests, such vulnerability can be an important part of growth: "And what is vulnerability? Just this: the ability to be wrong, to be foolish, to be weak and silly, to be an idiot. It is the ability to accept one's unworthiness, to accept one's vanity for what it is. It's the ability to be whatever and whoever you are—recognizing that you, like the world, like the earth, are fragile, and that in your fragility lies all possibility of growth and of death, and that the two are one and the same" (65).

3. *Deep listening entails asking for clarification.* Before we respond, we should clarify the speaker's message, to make sure that we've understood as fully as possible what s/he's saying.

4. *Deep listening entails frequent pauses and the ability to remain silent.* Sometimes it's best simply to listen, and not respond verbally (especially

if those responses would involve offering solutions, drawing analogies with our own experiences or those of others, or speaking without first self-reflecting).

5. *Deep listening enables us to challenge the ideas, not the speakers.* We can respectfully, but forthrightly, challenge desconocimientos, misunderstandings, and expressions of falsehoods and stereotypes about our own groups and other groups. When doing so, it is vital that we challenge the stereotypes/racism/ideologies/etc.—not the speaker herself.

Sources

Allen, Paula Gunn. *Off the Reservation: Reflections on Boundary-Busting, Border-Crossing Loose Canons.* Boston: Beacon P, 1998.

Anzaldúa, Gloria E. "now let us shift . . . the path of conocimiento . . . inner work, public acts." *this bridge we call home: radical visions for transformation.* Ed. Gloria E. Anzaldúa and AnaLouise Keating. New York: Routledge, 2002. 540–78.

Cervenak, Sarah J., Karina L. Cespedes, Caridad Souza, and Andrea Straub. "Imagining Differently: The Politics of Listening in a Feminist Classroom." *this bridge we call home: radical visions for transformation.* Ed. Gloria E. Anzaldúa and AnaLouise Keating. New York: Routledge, 2002. 341–56.

Hogue, Cynthia, Kim Parker, and Meredith Miller. "Talking the Talk and Walking the Walk: Ethical Pedagogy in the Multicultural Classroom." *Feminist Teacher* 12 (1998): 89.

Keating, AnaLouise. "Women of Color and Feminism: Twenty Years after *This Bridge Called My Back.*" Paper presented at New York University. Fall 2002.

WS/SOCI 5463. U.S. Women of Colors. Fall 2002.

Two Creation Stories

Genesis (Chapters 1–3)

Monotheistic: One very authoritative, masculine divine being
Top–down creation: (This God seems to do most of the work)
Humans and animals play small part in creation (Adam names the animals)
Humans' role primarily destructive ("the Fall")
Separation between the divine and the human
Separation between the human and nature
Separation between man (Adam) and woman (Eve)
Hierarchical arrangement: (1) God; (2) man; (3) woman; (4) animals
Begins with perfection; moves to state of exile and fallenness
Radical break between "good" and "evil"
Dualistic (human/divine; matter/spirit; humans/nature)
Emphasis on individuals
Domination of nature
Linear view of history

"Talk Concerning the First Beginning"

Multiple divine beings
Co-creation: Humans, animals, and divine play crucial, participatory
   roles in creation
Interrelatedness (kinship among all levels: animals, humans, land, spirits, etc.)
Holistic, nonbinary, and relational
   No radical break between material/spiritual; human/divine; etc.
   Shapeshifting; transformations

Harmony and balance
Nondualistic view of "good" and "evil"
 "Evil": illness, being out of harmony, rather than radically opposed to
 good
 Mistakes not necessarily permanent
Emphasis on communities, not individuals
Egalitarian
Nonhierarchical
Nonlinear

Sources

Allen, Paula Gunn. *The Sacred Hoop: Recovering the Feminine in American Indian Traditions*. Boston:
 Beacon, 1986.
Cajete, Gregory. *Look to the Mountain: An Ecology of Indigenous Education*. Skyland, NC: Kivaki
 P, 1994.
———. *Native Science: Natural Laws of Interdepencence*. Santa Fe: Clear Light Publishers, 2000.
Deloria, Jr., Vine. *God Is Red: A Native View of Religion*. Rev. ed. Goldon, CO: Fulcrum, 1994.

APPENDIX 4

Epistemologies of 'Whiteness'

I use the term *'whiteness'* to indicate a framework, an epistemology and ethics, that functions as an invisible norm which undergirds U.S. culture (educational systems, the media, etc.). I limit my remarks to 'whiteness' as represented and played out in U.S. culture; I do not equate 'whiteness' with people identified as 'white.' Rather, the relationship between 'white' people and 'whiteness' is contingent (Frye). I posit that we are all, in various ways, inscripted into this 'white'/supremacist framework. Because this framework functions to benefit 'white'-raced people, they can be more invested in it. This 'whiteness' intersects with certain versions of masculinity, economic status (middle- to upper-class), and colonization that has its roots in the Enlightenment.

Why 'Whiteness' as Framework?

"[S]ocietal norms and concepts of Americanness have developed in almost exclusively White political, social and cultural spheres."
(powell, "Our Private Obsession" 118–19)

"The result of this historical dominance is that the styles of thinking, acting, speaking, and behaving of the dominant group have become the socially correct or privileged ways of thinking, acting, speaking, and behaving. One of the main ways this happens is that the ways of the dominant group become universalized as measures of merit, hiring criteria, grading standards, predictors of success, correct grammar, appropriate behavior, and so forth."
(Scheurich 7)

"Whiteness fails to see itself as alien, as seen, as recognized. To see itself as seen, whiteness would have to deny the imperial epistemological and ontological base from which it sees what it wants (or has been shaped historically) to see. Whiteness refuses to risk finding itself in exile, in unfamiliar territory, an unmapped space of uncertainty in the form of both knowing and being. It refuses the difficult process of alienation and return, that is return to a different, antiwhiteness place of knowing and being. To refuse this process, whiteness denies its own potential to be Other (to be 'the not-same'), to see through the web of white meaning that it has spun. Whiteness refuses to transcend an economy of white discourse and action that creates the illusion of a social world of natural, immutable arrangements, arrangements that get axiological assignments (black = bad, white = good) from within a rigid system of white totalization."

(Yancy, "Introduction" 13)

"[W]e link whiteness with invisibility to highlight the normsetters' privilege of leaving their perspectives and practices unexcavated and unmarked, and of ignoring the perspectival nature of their perspectives. The way they see the world just is the way the world is, and the way they get around in the world just is the right way to get around. . . . Seeing whitely means participating in something much larger than oneself; it means tapping into a system of ideas and images that provides a kind of commonsense background for much of Western culture."

(Taylor 230)

"In educators' efforts to understand the forces that drive the curriculum and the purposes of Western education, modernist whiteness is a central player. . . . Objectivity and dominant articulations of masculinity as signs of stability and the highest expression of white achievement still work to construct everyday life and social relations at the end of the twentieth century. Because such dynamics have been naturalized and universalized, whiteness assumes an invisible power unlike previous forms of domination in human history. Such an invisible power can be deployed by those individuals and groups who are able to identify themselves within the boundaries of reason and to project irrationality, sensuality, and spontaneity on to the other."

(Kincheloe 164–65)

"Teachers need critical categories that probe the factual status of white, Western, androcentric epistemologies that will enable schools to be interrogated as sites engaged in producing and transmitting social practices that reproduce the linear, profit-motivated imperatives of the dominant culture, with its attendant institutional dehumanization."

(Giroux and McLaren 160)

Additional References. Aanerud, Chambers, Eze, Montag, Morrison, Thompson, Yancy.

Some Characteristics
of 'White' Epistemologies

NonRelational; Discrete Boundaries; Unspoken
Norm; Separates and Divides

In one of the few articles to focus specifically on epistemological 'whiteness,' Dwyer and Jones explore "two epistemological aspects of whiteness. The first of these, the social construction of whiteness, relies upon an essentialist and non-relational understanding of identity. Whiteness offers subjects who can claim it an opportunity to ignore the constitutive processes by which all identities are constructed. *In effacing their construction, 'white' people can paradoxically hover over social diversity just as they become the yardstick for its measurement.* This first moment is then linked to a second framing, a segmented spatialization that parallels the non-relational epistemology of white identities. This spatial epistemology relies upon discrete categorization of space—nation, public/private and neighbourhood [sic]—which provide significant discursive resources for the cohesion and maintenance of white identities. It also relies upon the ability to survey and navigate social space from a *position of authority.*"

(210, my emphasis)

"In everyday invocations of these categories [scale, boundaries, extensivity], both white and Other subjects reify social space, locating social subjects and attributing characteristics to places. This process of *categorical naturalization* is the spatial correlative of whiteness's non-relational social epistemology. In its solidification, it underwrites private property and the construction and orderly maintenance of segmented social spaces, from gated communities to redlined districts, from nature 'preserves' . . . to office towers. . . . Simultaneously marking and making difference by bounding white and Other in their respective places, this racialized geography has been reproduced on and through the built environment throughout American history."

(212–13, my emphasis)

"To find, to own, to catalogue, to encyclopedize, to collect, to measure: to register and to *textualize* became the driving sociopsychological, philosophical and epistemological impulse coextensive with 'colonial desire' and most expressly and meaningfully effected in the trade with African human beings. Or, to give a more graphic example: in John Locke's collected works, everything from the technology of olive oil presses through the history of Western navigation through the appropriate handling of slave property to the proper education of the young bourgeois figures, rather democratically, on the same level of self-empowered, obsessive interest in knowledge as ownership. For the white European (male) mind, knowledge and self-possession thus became inextricably articulated as one."

(Broeck 826, her emphasis)

Additional References. Broeck, Chambers, Dyer, Hughes, Myers, Sandoval.

Overreliance on Reason

"[A] dominant impulse of whiteness took shape around the notion of rationality of the European Enlightenment, with its privileged construction of a transcendental white, male, rational subject who operated at the recesses of power while at the same time giving every indication that he escaped the confines of time and space. In this context whiteness was naturalized as a universal entity that operated as more than a mere ethnic positionality emerging from a particular time, the late seventeenth and eighteenth centuries, and a specific space, Western Europe. Reason in this historical configuration is whitened and human nature itself is grounded upon this reasoning capacity. Lost in the defining process is the socially constructed nature of reason itself, not to mention its emergence as a signifier of whiteness. Thus, in its rationalistic womb, whiteness begins to establish itself as a norm that represents an *authoritative, delimited, and hierarchical* mode of thought. In the emerging colonial contexts in which Whites increasingly would find themselves in the decades and centuries following the Enlightenment, the encounter with nonwhiteness would be framed in rationalistic terms—whiteness representing orderliness, rationality, and self-control and nonwhiteness indicating chaos, irrationality, violence, and the breakdown of self-regulation. Rationality emerged as the conceptual base around which civilization and savagery could be delineated."

(Kincheloe and Steinberg 5, my emphasis)

Additional References. Dyer, Kincheloe, Leonardo, Sartwell.

Dualisms; Binary-Oppositional Thinking; Hierarchies

"As a scientific construct, whiteness privileges mind over body; intellectual over experiential ways of knowing; and mental abstractions over passion, bodily sensations, and tactile understanding."

(Kincheloe and Steinberg 5)

"Like many Indians and Mexicans, I did not deem my psychic experiences real. I denied their occurrences and let my inner senses atrophy. I allowed white rationality to tell me that the existence of the 'other world' was mere pagan superstition. I accepted their reality, the 'official' reality of the rational, reasoning mode which is connected with external reality, the upper world, and is considered the most developed consciousness—the consciousness of duality."

(Anzaldúa 36–37)

"The reasons for the false opposition [between "Americans" and "Chicanos" in student papers] have to do, again, with the extent to which white American reasoning (and the objective observer of Western epistemology) is predicated on the ability to divide abstractions into oppositional pairs: to think of "American" as an ever-shifting multiplicity of people resists oppositional thought."

(Stockton 172)

Additional References. Leonardo, Montag, Sandoval, Sartwell, Yancy.

Monolithic; Single-Voiced; Authoritative

Drawing on writings by Barthes and Fanon, Chela Sandoval explains that one of the ways the rhetoric of 'whiteness'/supremacism works is through "Identification" in which "perceptions of otherness are 'reduced to sameness' " (91).

Denies/ignores differences. "When enacting this [rhetorical strategy], consciousness must draw itself up, comfort, and 'identify' itself (or, as Fanon writes, constitute itself as 'Human') through a comparing and weighing operation that seeks to equate all varying differences with itself, the better then to either brush them aside as unimportant or to assimilate them." (Sandoval 90)

Additional References. Broeck, Frye, Montag, Sartwell, Taylor, Yancy.

Neutral, Distanced Objectivity; "Examinability" / Detachment; Ahistorical

"[E]xamination expresses a desire for the separation of examiner and examined that is itself an acknowledgement that the two are in fact connected. Examinability, in other words, is a device of disconnection: it presupposes the denegation of contexts—that is, of a history—in which the relatedness of the subjects and objects of examination becomes apparent. The act of examining . . . is a device for denying the contexts that join people. . . . [S]uch disconnectedness, as the denial of historical context, is what defines both the familiarity of the 'everyday' and the strangeness of the 'exotic.' "

(Chambers 194–95)

"Our neutral, neutralized, neutralizing, and neutered standards of beauty parade around in the robes of the universal. The calm, static, arid Western beauty that arises in the Renaissance, the ideal of a perfect balance and composure that makes possible a Kantian universalization of judgment of taste is, again, characteristic of our construction of ourselves. That is the *concealed* function of *our* arts in *our* construction of whiteness (a construction humorous in its bleakness), and it has, because it is white, to conceal itself *as* a communal construction."

(Sartwell 145, his emphasis)

Additional References. Baldwin, Dwyer and Jones, Kincheloe and Steinberg, Ratcliffe; Sandoval, Taylor, Yancy.

Self-Enclosed Hyper-Individualism

This extreme individualism defines the self in non-relational terms; it should not be confused with self-determination, personal agency, and self-respect.

"The European definition and exaltation of the individual were adopted in part to distinguish White Europeans from African, Indian and other non-White peoples that organized their society around non-individualistic norms. The construction of the individual, then, was neither natural nor race neutral, but was part of the racing process."

(powell, "Our Private Obsession" 111)

" 'hyperindividualism' is the 'major cultural distinction' of white Anglo-Saxon Protestants; it delimits interpersonal and social relationships and structures its concepts of 'self' as individual and "relationships" as contracts between individuals."

(Myser 4)

"[T]he category of the individual is the key to white hegemony, that is, to the unexaminedness of the degrees and divisions in whiteness (its own forms of nonwhiteness). The reason is that whiteness's indivisibility (as a function of the pluralization of the other) can be maintained only through the production of an invisibility that depends on atomizing whiteness (as a function of the homogenization of others)."

(Chambers 192)

Additional References. Goldberg; Mahoney; Myser; powell, "Disrupting Individualism;" Scheurich.

References

In the following list, an asterisk marks sources that I've found especially useful.

Aanerud, Rebecca. "Fictions of Whiteness: Speaking the Names of Whiteness in U.S. Literature." *Displacing Whiteness: Essays in Social and Cultural Criticism.* Ed. Ruth Frankenberg. Durham: Duke UP, 1997. 35–59.

Anzaldúa, Gloria. *Borderlands/La Frontera: The New Mestiza.* San Francisco: Spinsters/Aunt Lute, 1987.

*Baldwin, James. *Price of the Ticket: Collected Nonfiction 1948–1985.* New York: St. Martin's, 1985.

Broeck, Sabine. "When Light Becomes White: Reading Enlightenment through Jamaica Kincaid's Writing." *Callaloo* 25.3 (2002): 821–43.

*Chambers, Ross. "The Unexamined." *Whiteness: A Critical Reader.* Ed. Mike Hall. New York: New York UP, 1997. 187–203.

Delgado, Richard and Jean Stefancic, eds. *Critical White Studies: Looking Behind the Mirror.* Philadelphia: Temple UP, 1997.

Doane, Ashley W. and Eduardo Bonilla-Silva, eds. *White Out: The Continuing Significance of Racism.* New York: Routledge, 2003.

*Dwyer, Owen J. and John Paul Jones III. "White Socio-Spatial Epistemology." *Social and Cultural Geography* 1 (2000): 209–22.

Dyer, Richard. "White." *Screen: The Journal for Education in Film and Television* 29 (1988): 44–64.

————. *White*. New York: Routledge, 1997.

Dyson, Michael Eric. "The Labor of Whiteness, the Whiteness of Labor." *Race, Identity, and Citizenship*. Ed. Rudolfo D. Torres, Louis F. Mirón, and Jonathan Xavier Inda. Malden, MA: Blackwell Publishers, 1999. 219–24.

Eze, Emmanuel Chukwudi. *Achieving Our Humanity: The Idea of the Postracial Future*. New York: Routledge, 2001.

Flagg, Barbara J. " 'Was Blind, but Now I See': White Race Consciousness and the Requirement of Discriminatory Intent." *Critical White Studies: Looking Behind the Mirror*. Ed. Richard Delgado and Jean Stefancic. Philadelphia: Temple UP, 1997. 629–31.

Frankenberg, Ruth, ed. *Displacing Whiteness: Essays in Social and Cultural Criticism*. Durham: Duke UP, 1997.

Frye, Marilyn. "White Woman Feminist." *Willful Virgin: Essays in Feminism, 1976–1992*. Freedom, CA: Crossing, 1992. 147–69.

Giroux, Henry. "Rewriting the Discourse of Racial Identity: Towards a Pedagogy and Politics of Whiteness." *Harvard Educational Review* 67.2 (1997).

————. "White Squall: Resistance and the Pedagogy of Whiteness." *Cultural Studies* 11 (1997): 376–89.

Goldberg, David. *Racist Culture: Philosophy and the Politics of Meaning*. Cambridge: Blackwell, 1993.

Hall, Mike, ed. *Whiteness: A Critical Reader*. New York: New York UP, 1997.

Haney-López, Ian F. *White By Law: The Legal Construction of Race*. New York: New York UP, 1996.

Hughes, Langston. *The Ways of White Folks*. 1933. New York: Vintage, 1971.

Hurtado, Aída. "The Trickster's Play: Whiteness in the Subordination and Liberation Process." *Race, Identity, and Citizenship*. Ed. Rudolfo D. Torres, Louis F. Mirón, and Jonathan Xavier Inda. Malden, MA: Blackwell Publishers, 1999. 225–43.

Keating, AnaLouise. "Exposing 'Whiteness,' Unreading 'Race': (De)Racialized Reading Tactics in the Classroom." *Reading Sites: Gender, Race, Class, Ethnicity, and Sexual Orientation*. Ed. Elizabeth Flynn and Patsy Schweickart. New York: MLA, 2004. 314–42.

————. "Interrogating 'Whiteness,' (De)Constructing 'Race.' " *College English* 57 (1995): 901–18.

Kincheloe, Joe L. "The Struggle to Define and Reinvent Whiteness: A Pedagogical Analysis." *College Literature* 26 (Fall 1999): 162–94.

Kincheloe, Joe L. and Shirley R. Steinberg. "Addressing the Crisis of Whiteness." *White Reign: Deploying Whiteness in America*. Ed. Joe L. Kincheloe, Shirley R. Steinberg, Nelson M. Rodriguez, and Ronald E. Chennault. New York: St. Martin's P, 1998. 3–29.

Leonardo, Zeus. "The Souls of White Folk: Critical Pedagogy, Whiteness Studies, and Globalization Discourse." *Race, Ethnicity, and Education* 5.1 (2002): 29–50.

Lipsitz, George. *The Possessive Investment in Whiteness: How White People Profit from Identity Politics*. Philadelphia: Temple UP, 1998.

*————. "White 2K Problem." *Cultural Values* 4.2 (October 2000): 518–24.

Mahoney, Martha R. "The Social Construction of Whiteness." 1995. *Critical White Studies: Looking Behind the Mirror*. Ed. Richard Delgado and Jean Stefancic. Philadelphia: Temple UP, 1997. 330–33.

Montag, Warren. "The Universalization of Whiteness: Racism and Enlightenment." *Whiteness: A Critical Reader*. Ed. Mike Hill. New York: New York UP, 1997. 281–93.

Morrison, Toni. *Playing in the Dark: Whiteness and the Literary Imagination*. Cambridge: Harvard UP, 1992.

Myers, Kristen. "White Fright: Reproducing White Supremacy through Casual Discourse." *White Out: The Continuing Significance of Racism*. Ed. Ashley W. Doane and Eduardo Bonilla-Silva. New York: Routledge, 2003. 129–44.

Myser, Catherine. "Differences from Somewhere: The Normativity of Whiteness in Bioethics in the United States." *The American Journal of Bioethics* 3.2 (2003): 1–11.

Omi, Michael and Howard Winant. *Racial Formation in the United States from the 1960s to the 1980s.* Rev. ed. New York: Routledge, 1993.

powell, john a. "Disrupting Individualism and Distributive Remedies with Intersubjectivity and Empowerment: An Approach to Justice and Discourse." *University of Maryland Law Journal of Race, Religion, Gender and Class* 1 (2001): 1–23.

———. "Our Private Obsession, Our Public Sin: The "Racing" of American Society: Race Functioning as a Verb Before Signifying as a Noun." *Law and Inequality Journal* 15 (1997): 99–126.

Ratcliffe, Krista. "Eavesdropping as Rhetorical Tactic: History, Whiteness, and Rhetoric." *JAC: Journal of Advanced Composition* 20 (2000): 101–34.

Roediger, David, ed. *Black on White: Black Writers on What It Means to Be White.* New York: Schocken, 1998.

*Sandoval, Chela. "Theorizing White Consciousness for a Post-Empire World: Barthes, Fanon, and the Rhetoric of Love." *Displacing Whiteness: Essays in Social and Cultural Criticism.* Ed. Ruth Frankenberg. Durham: Duke UP, 1997. 87–106.

Sartwell, Crispin. *Act Like You Know: African-American Autobiography and White Identity.* Chicago: U of Chicago P, 1998.

Scheurich, James Joseph. "Toward a White Discourse on White Racism." *Educational Researcher* (1993): 5–10.

Scheurich, James J. and Michelle D. Young, "Coloring Epistemologies: Are Our Research Epistemologies Racially Biased?" *Educational Researcher* 26 (1997): 4–16.

Stockton, Sharon. " 'Blacks vs. Browns': Questioning the White Ground." *College English* 57 (1995): 166–81.

Taylor, Paul C. "Silence and Sympathy: Dewey's Whiteness." *What White Looks Like: African-American Philosophers on the Whiteness Question.* Ed. George Yancy. New York: Routledge, 2004. 227–41.

Thompson, Audrey. "Gentlemanly Orthodoxy: Critical Race Feminism, Whiteness Theory, and the *APA Manual.*" *Educational Theory* 54.1 (2004): 27–57.

Yancy, George. "Introduction: Fragments of a Social Ontology of Whiteness." *What White Looks Like: African-American Philosophers on the Whiteness Question.* Ed. George Yancy. New York: Routledge, 2004. 1–23.

Yancy, George, ed. *What White Looks Like: African-American Philosophers on the Whiteness Question.* New York: Routledge, 2004.

APPENDIX 5

Critics on Literary and Cultural Representations of 'Whiteness'

1. Invisible and unknown (Dyer, Keating)
2. Transparency (Flagg)
3. Unmarked, hidden superiority and dominance (Haney-López)
4. Unacknowledged standard or norm against which all others are measured and found lacking (Dyer, Frankenberg, Keating)
5. Invisible code of behavior (Frye)
6. "White" worldview (Dwyer and Jones, Frye, Keating, Roediger)
   Color-blind idealism
   Dominant/subordinate (hierarchical)
   Rigid boundaries
   Non-relational
7. Mystery, absence, death (hooks, Dyer)
8. Civilization, control, rationality (Dyer)
9. Repression (Dyer)
10. Freedom (Morrison)
11. Autonomy (Morrison)

Recent Analyses of 'Whiteness' and 'Race'

Dwyer, Owen J. and John Paul Jones III. "White Socio-Spatial Epistemology." *Social and Cultural Geography* 1 (2000): 209–22.

Dyer, Richard. "White." *Screen: The Journal for Education in Film and Television* 29 (1988): 44–64.

———. *White*. New York: Routledge, 1997.

Flagg, Barbara J. " 'Was Blind, but Now I See': White Race Consciousness and the Requirement of Discriminatory Intent." *Critical White Studies: Looking Behind the Mirror*. Ed. Richard Delgado and Jean Stefancic. Philadelphia: Temple UP, 1997. 629–31.

Frankenberg, Ruth, ed. *Displacing Whiteness: Essays in Social and Cultural Criticism*. Durham: Duke UP, 1997.

Frye, Marilyn. "White Woman Feminist." *Willful Virgin: Essays in Feminism, 1976–1992*. Freedom, CA: Crossing, 1992. 147–69.

Hall, Mike, ed. *Whiteness: A Critical Reader*. New York: New York UP, 1997.

Haney-López, Ian F. *White By Law: The Legal Construction of Race*. New York: New York UP, 1996.

Keating, AnaLouise. "Interrogating 'Whiteness,' (De)Constructing 'Race.' " *College English* 57 (1995): 901–18.

Morrison, Toni. *Playing in the Dark: Whiteness and the Literary Imagination*. Cambridge: Harvard UP, 1992.

Omi, Michael and Howard Winant. *Racial Formation in the United States from the 1960s to the 1980s*. Rev. ed. New York: Routledge, 1993.

Roediger, David, ed. *Black on White: Black Writers on What It Means to Be White*. New York: Schocken, 1998.

Webster, Yehudi O. *The Racialization of America*. New York: St. Martin's P, 1992.

APPENDIX 6

Sample Syllabi

This appendix contains the following undergraduate and graduate syllabi, arranged by ascending course numbers: Introduction to Literature, Introduction to Women's Studies, Studies in American Literature: U.S. Literature of Social Protest, U.S./American Literature I: Colonial through Early Federal Period, Literature of the Southwest, Feminist Epistemologies, U.S. Women of Colors, and Transgressive Identities: Queer Theories/Critical 'Race' Theories. To save paper, I have modified the syllabi by removing from each the section titled "Additional Course Policies," which includes policies on Disability Support Services, Academic Dishonesty, Class Visitors, Incompletes, Freewriting Activities, Cellphones, and Blackboard. I have also removed the sections on "Presuppositions for Dialogue" and "Listening," which you will find in appendices 1 and 2.

Introduction to Literature for English Majors and Minors

Dr. AnaLouise Keating
English 211

Course Description

This course focuses on analyzing, interpreting, and thinking about literature, literary merit, representation, and related issues. Through readings and discussions of a wide range of genres—including autobiography, drama, nonfiction prose, novels, poetry, and short stories—we will discuss strategies for reading, writing, thinking, and theorizing about literature.

Texts

American Poetry and Literacy Project, eds.: *101 Great American Poems* (GAP); Sandra Cisneros: *House on Mango Street*; Frederick Douglass: *Narrative of the Life of Frederick Douglass*; Don DeLillo: *White Noise*; Leslie Marmon Silko: *Ceremony*; Marie Harris, Kathleen Aguero, eds.: *An Ear to the Ground: An Anthology of Contemporary American Poetry* (EG); Anne Mazur, ed.: *America Street: A Multicultural Anthology of Stories* (AS); William Shakespeare: *Hamlet*.

Handouts (HO)

Jane Tompkins: excerpt from *Sensational Designs*; Nathaniel Hawthorne: "A Rill from the Town Pump;" Ralph Waldo Emerson: "Self-Reliance;" Audre Lorde: "Poetry Is Not a Luxury;" Richard McCann: "My Mother's Clothes;" Carmen Tafolla: "Frederica y Elfiria."

As You Read

For each assignment, I've included questions that we'll be discussing in class (marked *Think About*). Please take the time to consider these issues as you read, and take notes! Make sure you read each poem several times. Here are some additional questions you might ask yourself as you read: Do you like the material? Why or why not? What don't you understand? What questions do you have? Does it remind you of any other readings? How? In your opinion, does the text have "literary merit"? Why/not? *All readings must be completed by the date listed in the syllabus.*

E-Mail Discussion Group

I've set up a listserv discussion group specifically for this course. This discussion format will allow us to continue our conversations outside of class. If you have questions about the reading, or additional comments triggered by class discussions, send them to the listserv. At times, I might suggest a question for discussion. I will also use the list for announcements, reminders, unexpected changes in readings. *Please Note: To be eligible for a grade of C or higher, you must send at least one message to the listserv by October 1.*

Attendance Policy

Because much learning in this course occurs through dialogue, it's vital that you attend all classes. A significant portion of your grade is based on attendance. You should be aware that if you miss class you will lose attendance and quiz points. Also, if you miss more than five classes, I will deduct 10 points per absence.

Grades

Grades will be awarded on a point system. You can achieve the required number of points for each letter grade in a variety of ways:

Attendance. Each class is worth 10 points. (Points will be deducted for late arrivals and early departures. The specific number of points deducted will depend on how late you arrive and how early you leave.)

Quizzes. There will be a series of unannounced quizzes, based on the day's assigned readings and conducted at the beginning of the class period. Generally, I will draw quiz questions from the questions raised in the sections marked "*Think About*" and from questions I suggest on the listserv. All quizzes will be closed book; however, you may use your own notes. Each quiz will be worth at least 10 points. *Be on time to class! Quizzes cannot be made up.*

Discussion. Discussion is a vital part of the learning experience in this course. Literature is always open to multiple interpretations often based at least partially on each reader's experiences and beliefs; thus, it's important that we all respect each other's values, interpretations, and views. Because it's important that we all respect each other's needs, values, and views, our discussions will be guided by open-minded acceptance of divergent opinions. (Discussion is worth up to 200 points.)

Listserv. Each time you send a comment, question, or response to the listserv, you will receive at least 5 points. (Up to 45 additional points might be awarded based on the amount of time and thinking reflected.) You will not receive points if you simply respond to a previous comment by saying something like "I agree."

Interaction Papers. Typed, one- to two-page, single-spaced papers exploring one or more of the questions in the section marked "*Think About.*" As the word "interaction" suggests, you should actively respond to the readings assigned for each topic. In other words, don't just summarize the material; instead, use your own ideas, perspectives, and opinions as you interpret the texts. Generally, papers will be scored on a 25-point system [25–20: truly excellent; 20–17: very good; 15: good; and so on.] All scores will reflect both content and form. Papers must be turned in on the date of the assigned reading(s). *To be eligible for the grade of C or higher, you must complete at least one interaction paper.* I strongly recommend that you complete more than one. For additional guidelines see the last page of this syllabus.

Final Exam. Each student will select a poem from *101 Great American Poems* or *An Ear to the Ground* and lead a brief class dialogue on it; your dialogue should be wide-ranging, but be sure to explore issues related to literary merit. You must choose a poem not previously discussed in class. Poems must be selected by Tuesday of week 12. (Worth up to 75 points.)

Grade:	A	B	C	D	F
Points:	904+	904–804	803–703	702–603	602 or below

*Schedule of Class Meetings**

Week 1 T Introduction. Course Objectives. Narrative Perspective.

Th *Read*: Bambara: "Raymond's Run" (AS 10–22); Mori: "Business at Eleven" (AS 104–10).

Think About: How would you describe the narrator, Squeaky, in "Raymond's Run"? How does her view of competition change by the end of the story? How does her view of Gretchen change by the end of the story? Why do you suppose Bambara's story is called "Raymond's Run" (rather than, say, "Squeaky's Run")? How does the title shape your interpretation of the story? How would you describe the narrator in Mori's story? How does the narrator's perspective shape your perspective on John? How does it shape your interpretation of the story? Compare/contrast the two stories. What commonalities do you find? How do you account for similarities and differences?

Week 2 T *Read*: Mohr: "The Wrong Lunch Line" (AS 32–39); Hughes: "Thank You, Ma'am" (AS 49–54).

Think About: How would you describe Yvette and Mildred? Whose perspective do we get in Mohr's story? How would you describe the conflict in Mohr's story? How do the girls react to Mrs. Ralston? Do you find their reaction surprising? Why/not? How does Hughes'

---

* The following is a *tentative* schedule. I might change readings, due dates, or assignments, so please call me if you miss class.

title shape your expectations and interpretation of the story? How would you describe Mrs. Jones? How would you describe Roger? Whose perspective do we get in this story? What point(s) do you suppose Hughes was trying to make? Compare/contrast these two stories both with each other and with other stories for this course. What commonalities do you find? How do you account for similarities and differences?

Th *Read*: Paley: "The Loudest Voice" (AS 40–48); Tafolla: "Frederica y Elfiria" (HO).

Think About: How would you describe the narrator in Paley's story? How does the narrator's perspective shape our interpretation of events? Note how Paley begins the story using present tense verbs: What's the impact of using the present (rather than the past) tense? Why does Shirley's mother object to Shirley's role in the Christmas play? How did the code-switching (the use of Spanish words and phrases) impact your interpretation of Tafolla's story? How would you describe Frederica and Elfiria? Whose perspective do we get in this story? How would the story be different if told from a different perspective? In what ways, if any, does Tafolla's story critique conventional gender roles? Compare/contrast these two stories both with each other and with other stories for this course. What commonalities do you find? How do you account for similarities and differences?

Week 3 T *Read*: McCann: "My Mother's Clothes" (HO); Jen: "The White Umbrella" (AS 122–33).

Think About: Note the doubled perspective (childhood described from adult perspective) in McCann's story. Do you find this doubling effective? Why/not? How would you describe the adult narrator? How does this adult narrator seem to view his younger self? In what ways, if any, does this story critique conventional gender roles? What do you make of the story's ending? How would you describe the narrator in Jen's story? How would you describe the narrator's relationship with her mother? How might this story be different if told from the mother's or piano teacher's perspectives? What distinctions do the narrator and her family make between themselves and "Americans"? What does the white umbrella come to symbolize?

Th *Read*: Emerson: "Self-Reliance" (HO).

Think About: How would you define Emerson's self-reliance? What does Emerson mean by the "self": Is he referring only to the isolated individual? Is he defining it as something larger? Why do you suppose this essay is such a "classic" of American literature? How would you define Emerson's prose style?

Week 4 T *Read*: "Preface," by William Lloyd Garrison (Douglass vii–xiii); "*Letter from Wendell Phillips, Esq.*" (Douglass xv–xvii); Douglass: *Narrative of the Life of Frederick Douglass* (1–5).

Think About: Why do you suppose Douglass began his autobiography with this preface and letter? What functions do they serve? How would you describe Douglass's tone in his autobiography? Why do you suppose that some of his contemporaries did not believe that Douglass really wrote this autobiography? Note the various ways Douglass critiques slavery. What types of arguments and literary devices does he use?

Th *Read*: Douglass: *Narrative of the Life of Frederick Douglass* (5–59).

Think About: What is the relationship of literacy to Douglass's quest for freedom? What impact do you suppose Douglass wants to make on his readers? How does he attempt to engage readers' sympathies? Is he effective? Does Douglass enact, redefine, or reject Emerson's concept of self-reliance? What similarities and differences do you find between Douglass and Emerson? How do you account for them? How does Douglass represent slavery? Why is Douglass's battle with Covey "the turning point in [his] career as a slave" (43)?

Week 5 T *Read*: Douglass: *Narrative of the Life of Frederick Douglass* (59–76).

Think About: Why do you suppose Douglass included the appendix? How does he represent Christianity in the *Narrative*? Do you find his appendix persuasive? Why do you suppose the *Narrative* has, in the past fifteen years or so, become such a classic? How would you describe Douglass's writing style?

Th *Read*: Shakespeare: *Hamlet* Acts I, II, and III (1–78).

Think About: How would you describe Hamlet? In what ways is he similar to and different from the speakers in Emerson's essay and Douglass's autobiography? How do you account for these differences and similarities? What advice do you imagine Emerson and Douglass would give to Hamlet?

Week 6 T *Read*: Shakespeare: *Hamlet* Acts IV and V (79–122).

Think About: Why do you suppose *Hamlet* and Shakespeare are such classics? What similarities and differences do you find between *Hamlet* and other texts for this course? How do you account for the similarities/ differences?

Th *Read*: Cisneros: *House on Mango Street* (1–34).

Think About: How would you describe the narrator in this text? Is it significant that Esperanza's name also means "hope"? Why/not?

Film Screening: Stuart Hall *Representation and the Media*.

Week 7 T *Read*: Cisneros: *House on Mango Street* (25–66).

Think About: How does Cisneros represent Chicana identity? How does Esperanza's perspective shape the stories she tells? How would you describe the other characters in *House*? How does the title shape your interpretation?

Th *Read*: Cisneros: *House on Mango Street* (66 to end).

Think About: Would you describe *House on Mango Street* as a novel? As a collection of short stories? As both? As neither? Why? What similarities and differences do you find between *House on Mango Street* and other texts for this course? How do you account for the similarities/ differences?

Week 8 T *Read*: Tompkins: from *Sensational Designs*; Hawthorne: "A Rill from the Town Pump" (HO).

Think About: Remember that Tompkins's focus is not Hawthorne. Rather, she uses Hawthorne as an example to illustrate her larger point: What *is* her larger point?

How, according to Tompkins, have classics been defined by the academy? What does she seem to think about this definition? How does she define a classic? Interact with Tompkin's argument. Do you agree or disagree with what she's saying? Why?/Why not?

Th *Film Screening: Falling Down.*

Week 9 T *Read*: DeLillo: *White Noise* ("Waves and Radiation," 1–105).

Think About: How would you describe Jack? Does he remind you of any other characters we've encountered this semester? What's the significance of Jack's position as head of "Hitler Studies"? How would you describe Jack's family? Do they seem realistic? How does DeLillo represent contemporary U.S. culture? Does he seem to be critical of it? If so, how? How would you describe DeLillo's writing style? How does the title of this section—"Waves and Radiation"—shape your interpretation of it? What similarities—if any—do you see between Delillo's critique and the critique in *Falling Down*?

Th *Read*: DeLillo: *White Noise* ("The Airborne Toxic Event," 109–63).

Think About: In what ways (if any) do Jack and his family change in this section? If so, how do you account for these changes? How does DeLillo represent contemporary U.S. culture? Does he seem to be critical of it? If so, how? How does the title of this section—"The Airborne Toxic Event"—shape your interpretation of it?

Week 10 T *Read*: DeLillo: *White Noise* ("Dylarama," 167–326).

Think About: How would you describe Jack's quest? How does Jack change in the course of this novel? How does the title of this section—"Dylarama"—shape your interpretation? How does the novel's title shape your interpretation? Is *White Noise* an effective title? Why/not? In what ways (if any) does DeLillo seem to be critiquing contemporary U.S. society? What's the significance of Jack's fear of death?

Th *Read*: Silko: *Ceremony* (pages to be announced in class and on listserv).

Think About: How would you describe Tayo at the outset of this novel? How does Tayo's identity as a mixed-blood impact him? How would you describe his friends? How did their participation in World War II affect them? What connections do you see between the poetry and prose sections? Compare/contrast Silko's view of nature with Emerson's and DeLillo's. What commonalities, if any, do you find? How do you account for similarities and differences?

Week 11 T *Read*: Silko: *Ceremony* (pages to be announced in class and on listserv).

Think About: How does Silko depict the dominant society? Compare/contrast her representations with those in *White Noise* and *Falling Down*. Do you find one more effective than the others? Why/not? How would you describe Tayo's quest? Compare/contrast Tayo's quest with Jack's quest in *White Noise*: What similarities and differences do you find, and how do you account for them?

Th *Read*: Silko: *Ceremony* (finish reading the novel for today).

Think About: How does Tayo change during the course of the novel? What roles do women play in his transformation? In what ways does the novel itself function as a healing ceremony? Paul Lauter described *Ceremony* as a text that's "an actor for change." Does this seem like an applicable description? Why/not? Why do you suppose *Ceremony* has, in many ways, become a literary classic?

Week 12 T *Due*: The title of the poem you'll be discussing for your final.

Read: Lorde: "Poetry Is Not a Luxury" (HO); Rutsala: "Words" (EG 257–58); Tsui: "Don't Let Them Chip Away at Our Language" (EG 286–88); Moore: "Poetry" (GAP 65–66); MacLeish: "Ars Poetica" (GAP 72–73); Harper: "Songs for the People" (GAP 28–29).

Think About: How do today's writers depict language and/or poetry? What similarities and differences do you find? How do you account for these similarities and differences? Which views (if any) do you find most applicable to your own view of language? Why? Why does Lorde believe that poetry is not a luxury? What does she think it can do? How would you describe Lorde's writing style? What might Moore mean by "the genuine"?

Th *Read*: Robinson: "Richard Cory" (GAP 39); Dunbar: "We Wear the Mask" (GAP 43); Cullen: "Incident" (GAP 78–79); alurista: "southwestern trek in four part harmony" (EG 8–10); Baber: "handicaps" (EG 23); Baca: "Ese Chicano" (EG 24); Becker: "Dangers" (EG 28–29); Cisneros: "His Story" (EG 68–69).

Think About: How do today's poems depict love, sex, men, women, and the relationships between and among them? What similarities and differences do you find? How do you account for these similarities and differences? What's the impact of Cisneros's title on your interpretation of her poem?

Week 13 T *Read*: Clarke: "Of Althea And Flaxie" (EG 71–72); Clausen: "Sestina, Winchell's Donut House" (EG 73–74); Cooper: "Being Aware" (EG 77–78); Knight: "Circling the Daughter" (EG 155); Lewisohn: "For My Wife" (EG 164–65).

Think About: How do today's poems depict love, sex, men, women, and the relationships between and among them? What similarities and differences do you find? How do you account for these similarities and differences?

Th *Read*: Emerson: "The Snow-Storm" (GAP 5); Whitman: "A Noiseless Patient Spider" (GAP 24); Excerpt from "Song of Myself" (GAP 25–26); Allen: "Taku Skanskan" (EG 4–5); Brucac: "Prayer" (EG 46); Harjo: "Grace" (EG 115); Hogan: "Winter" (EG 130–31); Lyon: "The Foot-Washing" (EG 171); Ortiz: "Canyon de Chelly" (EG 218–19).

Think About: How do today's poems depict nature and/or the divine and/or spirituality-religion? How do you account for similarities and differences? How do they depict relationships between human beings and the outer world? Which depictions do you find most effective? Why? How would you describe Whitman's speaker?

Week 14 T *Read*: Hughes: "The Negro Speaks of Rivers" (GAP 77–78); Allen: "Molly Brant, Iroquois Matron, Speaks" (EG 5–7); Antler: "Bedrock Mortar Full Moon

Illumination" (GAP 14); Cornish: "Fannie Lou Hamer" (EG 82–83); Espada: "Where the Disappeared Would Dance" (EG 102–03); Hobson: "Central Highlands, Viet Nam, 1968" (GAP 125–26); Kaneko: "Wild Light" (EG 141–42); Mitsui: "Destination: Tule Lake Relocation Center" (194).

Think About: How do today's poems work with history, memory, and tradition? Do you see any similarities and differences?

Th Class presentations.

Week 15 T   Class presentations.

Th Class presentations.

Week 16 T   Class presentations.

★   ★   ★

Introduction to Women's Studies

Instructor: Dr. AnaLouise Keating
Contact Information:
Office Hours:

There are distinctions, certainly, about the way people are treated, but I want to reclaim the power to move through categories, so that I do not have to stay fixed in any one place.

Rayna Green

The world's definitions are one thing and the life one actually lives is quite another. One cannot allow oneself . . . to live according to the world's definitions: one must find a way, perpetually, to be stronger and better than that.

James Baldwin

Course Description

This course offers an introduction to Women's Studies as an interdisciplinary field. Focusing primarily on the United States, we'll use a variety of sources to explore the ways gender and other categories shape human identities and experiences and influence our perceptions, thinking, and actions.

Course Goals

Students who successfully complete this course will obtain the following: An increased understanding of the roles gender and other systems of difference play in shaping identities and lives; increased understanding of women's studies as an interdisciplinary field; increased knowledge of key issues, texts, and trends in women's studies; increased knowledge of the history of women and feminism/womanism in the United States; increased understanding of the various forms feminism/womanism takes; increased ability to read and think critically, and to express insights orally and in writing; increasingly nuanced understanding of the ways commonalities and differences work together; increased familiarity and skill with electronic technologies like the Internet; increased familiarity with the Woman's Collection at TWU's Blagg-Huey Library; increased ability to think relationally.

Required Texts and Supplies

Sisters in Spirit: Haudenosaunee (Iroquois) Influence on Early American Feminists by Sally Roesch Wagner
Kindred by Octavia Butler
Listen Up. Voices from the Next Generation of Feminists, ed. Barbara Findlen
3×5 inch Index Cards
Course Packet (CP)

As You Read

Please read carefully, and take notes! Here are some questions you might ask yourself as you read: Do you like the material? Why or why not? What don't you understand? What questions do you have? Does it remind you of any other readings? How? If there are words you don't understand, please look them up. *All readings must be completed by class time on the date listed in the syllabus.*

Attendance Policy

Discussion plays a major role in this course, and it's vital that you attend all classes. Consequently, a significant portion of your grade is based on attendance. If you miss class you'll lose points for both attendance and response cards. You are "allowed" four absences. I will deduct points for any additional absences; these deductions will significantly affect your grade. (See below.)

Grades

Grades will be awarded on a point system. This system gives students more flexibility and accountability. *It's your responsibility to keep track of your number of points.* You can achieve the required number of points for each letter grade in a variety of ways:

Attendance. Each class is worth 10 points. (Points will be deducted for late arrivals and early departures. The specific number of points deducted will depend on how late you arrive and how early you leave.) If you miss five, six, or seven classes, I will deduct 20 points for each class missed and you will be unable to receive any grade higher than a B. If you miss more than seven classes, I will deduct even more points (30 points for each additional class missed) and you will be unable to receive any grade higher than a C.

Response Cards. For each class on days when we have reading assignments you must bring index cards with at least one question concerning *each* of the readings assigned for that day. Please use a separate card for each reading. At the top of the card, put the reading's author and title. Write your question(s) on the lines under the card. Put your name on the other side of the card. I encourage you to take these cards seriously: They'll play the major role in shaping class discussion. (In other words, boring questions/comments = boring class time.) Each card is worth up to 5 points. (I reserve the right to give additional points for exceptionally insightful responses.) *Cards are due* before *I take attendance.* I do not accept late response cards.

Participation. Participation entails both discussion and engaged listening. We'll explore a variety of issues, some of them quite controversial. Thus, it's important that we all listen and speak respectfully. We do not need to agree with each other. As June Jordan states, "Agreement is not the point. Mutual respect that can accommodate genuine disagreement is the civilized point of intellectual exchange, particularly in a political context." Personal opinions and anecdotes play a role in class dialogues but do not substitute for discussions of the assigned readings. Discussion takes two forms: in-class dialogue and electronic dialogue on our listserv (described below). (Participation is worth up to 100 points.)

Campus Presentations. This semester, there will (probably) be a variety of presentations that in some way deal with issues we're exploring in this course. (For example, Deborah Miranda will give a lecture in March.) I'll announce additional presentations in class and on the listserv. If you attend a presentation and—within one week of the event—send a summary/analysis (approximately four well-developed paragraphs) to the listserv, you will receive up to 50 points.

Interaction Papers. Typed, two- to four-page, double-spaced papers that interact with one of more of the assigned readings. As the word "interaction" suggests, you should actively analyze and respond to the reading(s) you discuss. In other words, don't just summarize the material; instead, use your own ideas, perspectives, and opinions to develop a thesis statement (your main point) that will shape your essay. You are required to turn in one interaction paper, on your final project essay from *Listen Up*; it will be due on the day you present your project. Additional interaction papers are optional; they must be turned in on the date of the reading(s) or up to one week after the specific reading. Scores will reflect both content and form. To receive credit, you must interact with the reading. For additional guidelines see below. (Each paper is worth up to 30 points—or even more if exceptional.)

Library Projects. During the semester, we'll work on two projects in the library. See the final pages of this syllabus for copies of the assignments. (Project #1 is worth up to 100 points. Project #2 is worth up to 150 points.)

Final Project. Class Presentations. Working in groups of two or three, you'll select a reading from *Listen Up*, lead a brief class discussion, and write an interaction paper on it. For additional information, see page 9. Readings must be selected by Week 6. (Worth up to 100 points.)

E-Mail Discussion Group

I've set up a listserv discussion group specifically for this course. This discussion format will allow us to continue our conversations outside of class. If you have questions about the reading, additional comments triggered by class discussions, or want to make sure we discuss a specific issue during class, write to the listserv. At times, I might suggest a question for discussion or specific readings that I want to make sure we discuss in class. I will also use the list for announcements, reminders, unexpected changes in readings, etc. After you give me your e-mail address, I will subscribe you to the list. If you don't already have an e-mail account, please set one up before the third class. Please make sure to include your name on all postings.

Grade:	A	B	C	D	F
Points:	913+	917–816	815–714	713–612	Below 612

Library Project #1: Cookbooks

Select a cookbook that interests you from the Woman's Collection (TWU's Blagg-Huey Library). There are many interesting and rather unique cookbooks, so spend some time searching for a cookbook that intrigues you (or, at the very least, that you don't find extremely boring!). *Please note*: Not just any cookbook will work (although many will). You need to select a cookbook that can be analyzed for gender and other cultural clues. Once you've found "your" cookbook, read it. (You don't need to read all the recipes, but please do read the introductory material, the section introductions, and at least some of the recipe introductions.) As you read, look for gender clues in the illustrations, the introductory text, and the recipes themselves. Finally, focusing on gender (women and/or men) issues, develop a thesis statement and write a two- to three-page (typed, double-spaced) analysis/evaluation. Give specific examples from the cookbook to illustrate your points.

This project could take various forms, depending on what you select as your focus:

- You could select a cookbook from the eighteenth, nineteenth, or early twentieth century, and explore the ways that women's roles and identities seem to have changed (and, perhaps, the ways they have not changed) since that time. Give specific examples to illustrate the ways women were represented at the time the cookbook was written.

- You could select a regional cookbook and explore how women from that region are "particularized" (that is: how are these regional women defined and represented? what makes them different from women in other parts of the United States?)

- You could select a cookbook written during World War II and focus on patriotism: How is the United States represented during this time? How are women and their patriotic duty represented?

- You could select a cookbook targeted at men. When I was browsing the online catalog, I found a 1949 book called *Cookbook for Men Who Like to Cook*. How does this book, published in 1949, represent "men who like to cook" and, more generally, how does it represent men? In what ways are these 1949 gender roles similar to today's male gender roles? In what ways are they different? Does the author somehow try to justify the fact that some men (in the 1940s) like to cook?

- There are many other books targeted at men, children, or other specific groups of potential cooks. You could select one of these books and investigate issues related to gender roles. Yet another possibility: If you're feeling extremely energetic, you could select two cookbooks and compare them, focusing on a specific item (like female gender).

As you can see from these examples, the point is not to analyze the recipes themselves. Nor is the point to give a "book report" summary of the text. Rather, you must read the books searching for gender clues and write an analysis. We're treating these cookbooks as "material culture"— evidence about gender roles. For additional support, look at the two chapters by Sherrie A. Inness: "Paradise Pudding, Peach Fluff, and Prune Perfection: Dainty Dishes and the Construction of Femininity" (CP 112–23) and " 'Fearsome Dishes': International Cooking and Orientalism between the Wars" (CP 99–111).

You'll need a title for your essay. In your introduction, please include the following information: book title, author's full name, and publication date, as well as other introductory information and your thesis statement (which must be *underlined*).

Library Project #2: Feminist Periodicals

This assignment requires you to select a periodical from the Woman's Collection (TWU's Blagg-Huey Library), read/skim a number of issues, and prepare a typed, double-spaced analysis/evaluation, focusing on its feminist (or non-feminist) dimensions. In order to complete this assignment successfully, you will need to arrive at your own definition of feminism. Please use class discussions and course readings to develop, illustrate, and support your definition of feminism. You do not need to write this paper as a standard essay. Instead, structure your paper according to this format:

Periodical's Background. At least one paragraph providing background information about the periodical. You will want to include information about: (1) how often the periodical is published; (2) how long the periodical has been in existence; (3) the editor/publisher and what are her/his qualifications; (4) the types of writing included (scholarly? personal narratives? self-help articles? etc.?); (5) the types of authority cited by article authors (Do the authors refer primarily to professionals? Do they refer to real-life "ordinary" people?) *Note*: Give specific examples to illustrate your information.

Periodical Purpose/Audience. At least one paragraph explaining the periodical's purpose and targeted audience. Be as specific as possible and use data like editorials, article content, table of contents, and advertisements to support your assertions.

Shifts in Purpose/Audience. At least one paragraph exploring how the purpose and/or targeted audience have changed over the years. Please give examples. ("Advertisements from the July 1967 issue focus on food-related items, while advertisements from the July 2002 issue . . . .") Be as specific as possible and use data like editorials, article content, table of contents, and advertisements to support your assertions. Or, if the purpose and targeted audience haven't changed, explain and give examples supporting your position.

Feminist Analysis. At least two paragraphs analyzing the journal's (non)feminist dimensions. In the first paragraph, give your definition of feminism. Make sure to explain why you've selected/developed your specific definition. *Note:* You cannot simply define "feminism" as "anything related to women." Your definition must be more nuanced. In paragraphs 2 and 3, use your definition to analyze the (non)feminist dimensions of your periodical. Here are some of the questions that might help: What types of issues does the periodical explore? Are these issues of interest to feminists? Why/not? Give specific examples for support. (Be as specific as possible and use data like editorials, article content, table of contents, and advertisements to support your assertions.)

Usefulness. At least one paragraph discussing the following: Would you recommend that your classmates read this journal? Please explain why or why not. For instance, you might discuss whether the journal explores issues we've discussed in class and, if so, which issues.

Grades. This project is worth up to 150 points. As usual, grades will be based on both content and form. Grades will also be based on the amount of work you put into the paper.

Final Project

The final project focuses on *Listen Up* and requires students to lead a class discussion and write an interaction paper on one of the essays in *Listen Up*. During Week 6, you'll choose the three essays that you would most like to use for your final project. We will assign groups of 2 to 3 students, based on student preferences. You may also feel free to select your own group; if you do so, turn in a single list of preferences for the entire group. Working with your group, you'll design a set of discussion questions.

It's crucial that all members participate equally in designing the questions. If one of your group members does not participate, please let the instructors know.

During the 15-minute presentation, you'll lead a discussion, asking your classmates (who will have read the essay ahead of time) questions about your text. Also at the time of your discussion, you will turn in an interaction paper on your selected essay.

We will have class time to work with your group. Please make every effort to attend during these days. I also recommend that you research your author.

The final project includes:

1. A list of *at least ten questions* that you'll be discussing with the class. This list must be typed, double-spaced. Due at the time of your class presentation. (Each group will turn in one set of questions. Make sure to list all group members' names.)

2. *Interaction paper*: An essay, (at least three full, double-spaced, typed pages) that analyzes your selected text. Do not summarize your text. Instead, analyze it, focusing on issues we discussed in this class. For additional suggestions, see "Guidelines For Written Work" and below. Due at the time of your class presentation. *Please Note*: Each person must write an interaction paper.

Scores will be based on the following criteria:

- How much research, thought, and other forms of effort went into this project?
- Are you extremely familiar with your selected text and author? (Some of the authors have written other books, have well-known mothers, etc. It's important that you be aware of this information.)
- How well-organized and thought-provoking are your questions?
- Do some of your questions draw connections with other course readings and/or issues?
- Do your questions engage your classmates?
- How much time and thought did you put into writing your interaction paper?
- Does your interaction paper follow the guidelines?
- Have you learned much from your project?

I'll be glad to talk with you about your projects throughout the semester.

Schedule of Class Meetings[*]

Week 1 M Introduction: Definitions. Course expectations. Why Women's Studies?

W *Read*: Anne Sexton: "Cinderella" (CP 5–7); Ekua Omosupe: "In Magazines (I Found Specimens of the Beautiful)" (CP 26).

Due: Bring to class one or two advertisements picturing women (and men). Each student will give a brief report on the images, focusing especially on what they indicate about contemporary representations of women (and men).

Week 2 M No class (Martin Luther King Day).

W *Read*: Michael Gershman: "Kotex" (CP 1–4); Arlene Dahl: "Always Ask a Man" (CP 22–25).

Week 3 M *Read*: Sherrie A. Inness: "Paradise Pudding, Peach Fluff, and Prune Perfection: Dainty Dishes and the Construction of Femininity" (CP 112–23).

Tour of the Woman's Collection in the Blagg-Huey Library. We'll meet in our classroom and walk over together.

W *Meet* in the Woman's Collection (Blagg-Huey Library). Work on Library Project #1.

Week 4 M *Meet* in the Woman's Collection (Blagg-Huey Library). Work on Library Project #1.

Recommended: Inness: " 'Fearsome Dishes': International Cooking and Orientalism between the Wars" (CP 99–111)

W *Due*: Rough draft and tentative thesis statement for Library Project #1. They must be typed. *Note*: Please bring two copies of your thesis statement.

Week 5 M *Due*: Library Project #1.

Read: Audre Lorde: "Transformation of Silence Into Language and Action" (CP 96–98); Mitsuye Yamada: "Masks of Woman" (CP 79–80).

W *Read*: *Sisters in Spirit* 10–36; Winona LaDuke: "Indigenous Women" (CP 15–19).

Recommended: Louise Bernikow: "What Really Happened at Seneca Falls" (CP 8–12).

[*] The following is a *tentative* schedule; I might change readings, due dates, or assignments, so please e-mail or call me if you miss class.

Week 6 M *Read*: *Sisters in Spirit* 37–62; "Declaration of Sentiments" (CP 13–14).

Recommended: Bonnie Thornton Dill: "Fictive Kin, Paper Sons, and Compadrazgo: Women of Color and the Struggle for Family Survival" (CP 68–78).

W *Read*: *Sisters in Spirit* 63–98; Pat Mora: "Legal Alien" (CP 42).

Week 7 M *Read*: Patricia Lunneborg: "Abortion: A Positive Decision" (CP 54–63); Keesa Schreane: "appraising god's property" (CP 64–67).

Video: *From Danger to Dignity*.

Due: Final Project essay selection. Select the three essays in *Listen Up* that you would most like to use for your final project. For each selection please include the author's full name, the title, and the page numbers.

W *Read*: Susan Muaddi Darraj: "It's Not an Oxymoron: The Search for an Arab Feminism" (CP 34–42); Leslie Marmon Silko: "Yellow Woman and a Beauty of the Spirit" (CP 27–33).

Week 8 M *Read*: Vicki Sears: "Keeping Sacred Secrets" (CP 45–53); Bernadette García: "This World Is My Place" (CP 142).

W *Read*: Octavia Butler: *Kindred* (9–57).

Recommended: "Octavia Butler" (CP 81–85).

Week 9 M *Read*: Butler: *Kindred* (58–143).

W *Read*: *Kindred* (143–97).

Week 10 M *Read*: *Kindred* (198–264).

W *Meet* in the Woman's Collection (Blagg-Huey Library). Library Project #2.

Week 11 M *Meet* in the Woman's Collection (Blagg-Huey Library). Library Project #2.

W *Read*: "Abortions, Vacuum Cleaners . . ." (*Listen Up* 112–17).

Week 12 M *Due*: Library Project #2.

Read: susan jane gilman: "klaus barbie, and other dolls I'd like to see" (CP 86–89); meredith mcghan: "dancing toward redemption" (CP 90–95).

W   *Due*: Five discussion questions for your *Listen Up* essay.
Bring *Listen Up* to class. Work on group projects.

Week 13   M   Bring *Listen Up* to class. Work on group projects.

W   Bring *Listen Up* to class. Work on group projects.

Week 14   M   *Read*: *Listen Up*: TBA. Student-Lead Discussions.

W   *Read*: *Listen Up*: TBA. Student-Lead Discussions.

Week 15   M   *Read*: *Listen Up*: TBA. Student-Lead Discussions.

W   *Read*: *Listen Up*: TBA. Student-Lead Discussions.

Week 16   M   We meet from 1:30 to 3:30 p.m.

Read: *Listen Up*: TBA. Student-Lead Discussions.

★ ★ ★

Studies in American Literature: U.S. Literature
of Social Protest

Dr. AnaLouise Keating
English 273

> Language can construct understanding, language can assault, and
> language can exclude. Words have power.
>
> Mari J. Matsuda

Course Description

This course explores social protest in nineteenth- and twentieth-century
U.S. American literature from a variety of genres, including poetry,
short stories, autobiography, slave narrative, fiction, science fiction,
speeches, essays, and journal articles. Issues we'll be focusing on include
the following: In what ways have U.S. authors used the written and spoken
word to try to achieve social justice? How do they define social justice?
What strategies do they use in their attempts to achieve it? How
effective are these strategies? Are some strategies more effective than
others? Are some genres more effective than others? How does audience
shape rhetorical strategies? What are the intersections between art and politics? *Do* words have power? *Can* literature help us achieve social justice?

Required Texts

Frederick Douglass: *Narrative of the Life of Frederick Douglass, An American
Slave*

Maxine Hong Kingston: *The Woman Warrior*
Marge Piercy: *Woman on the Edge of Time*
Tomás Rivera: *Y no se lo trago la tierra/And The Earth Did Not Devour Him*
Roberto Santiago, ed.: *Boricuas: Influential Puerto Rican Writings*
Leslie Marmon Silko: *Ceremony*
Harriet Beecher Stowe: *Uncle Tom's Cabin*

Handouts (marked HO in syllabus)

Ralph Waldo Emerson: "Self-Reliance;" Martin Luther King: "I Have a Dream;" Malcolm X: "The Ballot or the Bullet;" Sandra Cisneros: "Little Miracles, Kept Promises" and "Salvador Late or Early."

Objectives

To become acquainted with and appreciate a variety of nineteenth- and twentieth-century U.S. literary texts; to explore the intersections between politics and art; to reexamine our conceptions of "American literature" and literary merit; to investigate issues related to social justice and literary activism; to develop powers of critical reading and thinking through in-class discussions and writing assignments; to become more familiar with the Internet; to develop cohesiveness as a classroom community; and to have fun.

As You Read

For each reading assignment, I've included questions and topics that we'll be discussing in class. (Marked *Think About.*) Please reflect on these issues as you read, and take notes! Here are some additional questions you might ask yourself as you read: Do you like the material? Why or why not? What don't you understand? Does it remind you of any other readings? How? Would you describe the text as social protest? Why/not? What strategies does the writer employ as s/he attempts to persuade readers? What specific institutions, practices, values, and/or beliefs does the author challenge? What kinds of social transformation does the author advocate? Please read each poem several times; look up words you don't know. *All readings must be completed by the date listed in the syllabus.* Always bring the assigned text(s) to class.

Discussion Group

I have set up a listserv discussion group specifically for this course. This discussion format will allow us to continue our conversations outside of class. If you have questions about the reading, additional comments triggered by class discussions, or want to make sure we discuss a specific issue or passage during class, write to the listserv. At times, I might suggest a question or topic that I want to make sure we talk about in class. I will also use the list for announcements, reminders, unexpected changes in readings, etc. Everyone must post at least one message to the listserv, and they must do so before the fourth week of class. Please make sure to include your name and e-mail address on all postings.

Attendance Policy

Because much learning in this course occurs through dialogue, it's vital that you attend all classes and a significant portion of your grade is based on attendance. If you miss class you'll lose attendance points, quiz points, and more. For specific penalties, see below.

Grades

Grades will be awarded on a point system. You can achieve the required number of points for each letter grade in a variety of ways.

Attendance. Each class is worth 10 points. Points will be deducted for late arrivals and early departures. You are "allowed" four absences. If you miss five to seven classes, you will lose 20 points for each additional class missed and you will be unable to receive any grade higher than a B. If you miss more than seven classes, you will lose 30 points for each additional class missed and you'll be unable to receive any grade higher than a C.

In-Class Essays. At least once a week, we'll begin class with a brief in-class essay based on the day's assigned readings. These essays are designed to measure how carefully you've read and thought about the material. Generally, I will draw essay questions from the questions raised in the sections marked *Think About* and from questions and issues I suggest in class or on the listserv. The essays will be closed book; however, you may use your own notes and your syllabus. Each essay will be worth up to 15 points, based on both content and form. (I reserve the right to give additional points for exceptionally insightful answers.) *Essays cannot be made up, so be on time!*

Discussion. Discussion is essential to the learning experience in this course. Because literature is always open to multiple interpretations often based at least partially on each reader's experiences and beliefs, it's important that we all respect each other's values, interpretations, and views. Personal opinions play a role in class dialogues but do not substitute for sustained effort to understand and interpret all course material and to use textual evidence to support your interpretations. (Discussion is worth up to 200 points.)

Listserv. Each time you send a comment, question, or response to the listserv, you will receive at least 5 points (unless you simply respond to a previous comment by saying something like "I agree.") Up to 45 additional points might be awarded based on the amount of time and thinking reflected.

Interaction Papers. Typed, one- to two-page, single-spaced essays analyzing the assigned readings. As the word "interaction" suggests, you should actively respond to the readings. In other words, don't just summarize the material; instead, use your own ideas, perspectives, and opinions as you interpret the text(s). Generally, papers will be scored on a 30-point system [30–29: truly excellent; 28–26: very good; 25–22: good; 21–19: average; and so on.] All scores will reflect both content and form. Papers must be turned in on the date of the assigned reading(s) or up to one week after a specific reading. You're required to turn in one interaction paper with your final project. Additional interaction papers are optional. For additional guidelines see the last page of this syllabus.

Final Project. Each student is required to select an excerpt from *Boricuas*, lead a brief class discussion, and write an interaction paper and self-reflective essay on it. For additional information, see below. Readings must be selected by 3/17. (But feel free to make your selection earlier.) Worth up to 150 points.

Grade Points

A 946+   A− 945–915   B+ 914–875   B 874–826   B− 825–796
C+ 795–776   C 775–736   C− 735–695   D+ 694–677   D 676–627
D− 626–597

Instructions for Final Projects

Working alone or in groups of two, you must select an excerpt from *Boricuas*, lead a class discussion, and write an interaction paper and a self-reflective essay on it. Readings must be selected by Week 9. (But feel free to make your selection earlier.) Worth up to 150 points.

All final projects must include the following components:

1. Project proposal, due Week 9 or earlier.
   Your proposal should be one typed, double-spaced paragraph that lists your selection (author, title, and page numbers) and briefly discusses why you chose that particular piece.
2. Class presentation.
   Your classmates will read your text ahead of time.
   During the presentation, you'll lead a class discussion, asking them questions about your text.
   You'll conclude by talking briefly about your text selection (why you chose it) and what you learned from your research.
3. List of *at least* ten questions that you'll be discussing with the class.
   This list must be typed, double-spaced.
   Due at the time of your class presentation.
4. Interaction paper.
   An essay (at least five double-spaced, typed pages) that analyzes your selected text.
   Do not summarize your text. Instead, analyze it, focusing on issues related to social protest and other issues we discussed in this class.
   For additional suggestions, see below.
   Due one week after your class presentation.
5. Brief self-reflective essay.
   An essay, at least three double-spaced, typed pages, that reflects on what you've learned about literature (especially literature as social protest) from preparing and presenting your text.
   Due one week after your class presentation.

Please Note: If you're doing your presentation with a classmate, you must each write an interaction paper and self-reflective essay.

Some issues to reflect on as you select your text and prepare your report:

- Would you define your selection as social protest? Why/not?
- If viewed as social protest, how effective is your text?
- What specific issues, practices, values, beliefs, and/or institutions does your author explore?
- What other readings from this course does this text remind you of?
- In what ways (if any) is your text unique?

Scoring for Final Projects. This project is worth up to 150 points. (I reserve the right to give additional points to incredibly exceptional projects.)

Scores will be based on the following criteria:

- How much research, thought, and other forms of effort went into this project?
- Are you extremely familiar with your selected text?
- Have you learned much from your project?
- Has your project enhanced your appreciation and knowledge of U.S. literature?
- Does your presentation engage your classmates?
- Does it provoke them to think differently about your specific text and about U.S. literature in general?
- How much time and thought did you put into writing your inter-action paper and your self-reflective essay?

Schedule of Class Meetings*

Week 1   T   Introduction. Course Objectives. What is "social protest"?

   F   *Read*: Emerson: "Self-Reliance" (HO).

   Think About: How would you define Emerson's self-reliance? What does Emerson mean by the "self": Is he referring only to the isolated individual, or is he defining it as something larger? How does Emerson view society? Is this essay relevant to today's readers? Why/not? Why do you suppose this essay is such a "classic" of U.S. American literature? How would you describe Emerson's prose style?

Week 2   T   *Read*: "Preface," by William Lloyd Garrison (Douglass 33–42); "Letter from Wendell Phillips, Esq." (Douglass 43–46); Douglass: *Narrative of the Life of Frederick Douglass* (47–63).

   Think About: Why do you suppose Douglass began his autobiography with this preface and letter? What functions do they serve? How would you describe Douglass's tone in his autobiography? Why do you suppose that some of his contemporaries did not believe that Douglass really wrote this autobiography?

---

* The following is a *tentative* schedule. I might change readings, due dates, or assignments, so please call or e-mail me if you miss class.

F *Read*: Douglass: *Narrative of the Life of Frederick Douglass* (65–100).

Think About: What is the relationship of literacy to Douglass's quest for freedom? What impact do you suppose Douglass wants to make on his readers? How does he attempt to engage readers' sympathies? Is he effective?

Week 3   T   *Read*: Douglass: *Narrative of the Life of Frederick Douglass* (101–59).

Think About: What similarities and differences do you find between Douglass and Emerson? How do you account for them? Does Douglass enact, redefine, or reject Emerson's concept of self-reliance? Why is Douglass's battle with Covey "the turning point in [his] career as a slave" (43)? Why do you suppose Douglass included the appendix? How does he represent Christianity in the *Narrative*? Do you find his appendix persuasive? Why do you suppose the *Narrative* has, in the past fifteen years or so, become such a classic? How would you describe Douglass's writing style?

F *Read*: Stowe: *Uncle Tom's Cabin* (41–161).

Think About: The original subtitle for this novel was "The Man Who Was a Thing." Is this an appropriate description of Tom? Why/not? How would you describe Tom? How do the chapter titles influence readers' expectations? In what ways does Stowe stereotype enslaved people? How does Stowe use Eliza's situation to comment on slavery? Do you find any similarities between Stowe's novel and Douglass's narrative?

Week 4   T   *Read*: Stowe: *Uncle Tom's Cabin* (162–304).

Think About: How would you describe St. Clare, Miss Ophelia, and Eva? How does Stowe use these characters to comment on slavery? How does Stowe critique religion? Why do you suppose this novel was so incredibly popular?

F *Read*: Stowe: *Uncle Tom's Cabin* (305–456).

Think About: How does Stowe use the interaction between Miss Ophelia and Topsy to comment on slavery? In what ways does Stowe hold the North, as well as the South, accountable for slavery? How would you describe the friendship between Tom and Eva? How does Stowe use the conversations in today's reading to critique slavery? Why do you suppose this novel has, until recently, rarely been studied in college courses?

Week 5   T   *Read*: Stowe: *Uncle Tom's Cabin* (457–629).

Think About: How does Stowe represent slave auctions? What points might she be making about slavery? In what ways are her arguments similar to Douglass's? In what ways are they different? How would you describe Simon Legree, Emmeline, Quimbo, and Sambo? How does Stowe use them to critique slavery? In what ways is Tom a martyr? Do the "Concluding Remarks" add to the novel? Why/not? Did reading this book change you? If so, how? In what ways can *Uncle Tom's Cabin* be described as a "story of salvation through motherly love"? Is *Uncle Tom's Cabin* still relevant today? Why/not? Why do you suppose Douglass's *Narrative* is taught more often than Stowe's novel? Did you find one more persuasive than the other? If so, why?

F   *Read*: King: "I Have a Dream" (HO); Dunbar: "We Wear the Mask" (HO); Malcolm X: "The Ballot or the Bullet" (HO).

Think About: How is "America" portrayed in King's speech? What dreams does King have for America? Is this speech still relevant to today? Why/not? What strategies does King employ to persuade his audience? What rhetorical strategies does Malcolm X employ in his speech? Malcolm X delivered "The Ballot or the Bullet" to a predominantly African-American meeting in Cleveland of the Congress of Racial Equality (CORE). How do you think this audience influenced Malcolm X's speech? Compare/contrast Malcolm X's language with King's.

Week 6   T   *Read*: Rivera: *And the Earth Did Not Devour Him* (83–117).

Think About: How would you describe the narrator/ protagonist? How would you describe the other characters? What's the effect of including multiple subjectivities (perspectives)? What's the significance of the title? How does it shape your interpretation of the text? How does the form—the bilingual text and way the narratives are laid out on the page—impact your interpretation? In what ways is this text different from more conventional novels? What does Rivera seem to say about migrant life? about religion? about education?

F *Read*: Rivera: *And the Earth Did Not Devour Him* (118–52); Cisneros: "Salvador Late or Early" (HO).

Think About: According to Teresa McKenna, "Tomás Rivera explores the psychological and political dilemmas of migrant life, the movement from place to place, and its effects on family, religious belief, and sense of self." Do you agree with this interpretation? Why/not? Would you describe *And the Earth Did Not Part* as a novel? a collection of short stories? both? neither? What impact (if any) does the text's unconventional format have on Rivera's social protest: Does it contribute to it? Detract from it? Not impact it at all? Would you describe Cisneros's short short story as social protest? Why/not? How would you describe Salvador? How does Cisneros describe herself in her short bio included on this page? What does this bio tell you about her?

Week 7 T *Read*: Sandra Cisneros: "Little Miracles, Kept Promises" (HO).

Think About: Pay attention to the diverse voices in Cisneros's story. What kinds of voices does she include? What types does she exclude? Do some voices seem to speak with more authority than others? Do any of the voices seem to represent Cisneros's perspective? Compare/contrast her use of multiple voices to Rivera's. How does Cisneros depict religion/spirituality? Would you describe her use of spirituality and myth as a form of social protest? Why/not?

F *Read*: Roberto Santiago: "Introduction" (*Boricuas* xiii–xxxiii); Sandra Maria Esteves: "Here" (*Boricuas* 3); José de Diego (*Boricuas* 25–26); Rosario Morales: "Double Allegiance" (*Boricuas* 74–76); Aurora Levins Morales: "Child of the Americas" (*Boricuas* 79–80); Pedro Pietri: "Puerto Rican Obituary" (*Boricuas* 117–26); Pedro Juan Soto: "Bayamaña" (*Boricuas* 243–44); Julia de Burgos: "I Became My Own Path" (*Boricuas* 255–56).

Think About: How does Santiago define Puerto Rican? In this introduction, Santiago describes the six sections of the anthology. Which sections seem especially interesting to you? Why do they seem interesting? How does Esteves define herself? How does she identify with her heritage? Who did you assume the "Gentlemen" addressed in the first line of de Diego's poem to be? What is de Diego protesting? How does he use history to develop his protest? What, according to Levins Morales, does it mean to be a "child of the americas"? To what two things does Rosario Morales feel allegiance? How does this "double allegiance" affect her? How does Pietri describe Puerto Ricans? How do they view America? How does Pietri seem to view America? What is the conflict in Soto's "Bayamaniña"? How is it resolved? What point(s) do you think Soto is trying to make? How does he make his points? What do you think de Burgos means when she says she became her "own path"? What path(s) is she rejecting?

Week 8   T   *Read*: Kingston: *Woman Warrior* (1–53).

Think About: What's the effect of beginning with "No Name Woman"? (That is, what expectations does it set up in your mind as a reader?) How does the aunt's story "branch into" Kingston's own life (10)? Is the aunt simply a victim figure? Or does she get some type of revenge? What does this chapter tell us about gender roles? What seems to be Kingston's perspective? What roles does story-telling play in *Woman Warrior*? What type of heroine is Fu Mu Lan? Compare/contrast her with the aunt in chapter 1. In what ways (if any) does Fu Mu Lan's story allow Kingston to redefine female identity and, more personally, her own identity?

F   *Read*: Kingston: *Woman Warrior* (53–109).

Think About: How would you describe Maxine's mother in this chapter? How does this chapter's depiction of the mother revise the depiction developed in the previous chapters? How does Maxine define herself in relation to her mother?

Week 9   T   *Read*: Kingston: *Woman Warrior* (110–209).

Think About: How would you describe Moon Orchid? Compare/contrast her with the other women we've encountered in *Woman Warrior*. What does it mean to be an "outlaw knot-maker" (163)? What points do you think

Kingston is making about silence and speech? How does Maxine define herself in relation to her community? How does she define herself in relation to her culture? Would you describe *Woman Warrior* as autobiography? as fiction? as both? as neither? Do the chapters have particular themes?

F *Due*: Final Project Proposal.

Read: Piercy: *Woman on the Edge of Time* (1–71).

Think About: Did you find the violence in the opening pages gratuitous, or is it used to make a larger point? How would you describe Connie, Dolly, and Lucretia? How does Piercy use these characters to critique gender roles? How is Piercy critiquing twentieth-century medicine? How is she critiquing twentieth-century gender categories and roles?

Week 10 T *Read*: Piercy: *Woman on the Edge of Time* (72–183).

Think About: How does Piercy critique twentieth-century nuclear families, heterosexual relationships, reproductive techniques, sexuality, childrearing practices, and economics? What else does she critique? How would you describe the twenty-first-century world she depicts?

F *Read*: Piercy: *Woman on the Edge of Time* (184–295).

Think About: How would you describe Gildina's world? What points do you think Piercy is trying to make about science and technology? What alternatives to twentieth-century versions of science/technology does she offer? What "war" is Connie fighting?

Week 11 T *Read*: Piercy: *Woman on the Edge of Time* (296–376).

Think About: How does Connie change in the course of this novel? Do you see these changes as positive? negative? neutral? Why? Did you find the novel's ending satisfying? Why/not? Pay attention to the last chapter. What errors do you find in this clinical report?

F *Read*: Silko: *Ceremony* (1–38).

Think About: How would you describe Tayo at the outset of this novel? How would you describe his friends? How did their participation in World War II affect them?

What connections do you see between the poetry and the prose sections?

Week 12 T *Read*: Silko: *Ceremony* (38–138).

Think About: Why do Leroy, Emo, Pinkie, and Harley tell stories about the war? Why do their stories disturb Tayo? How would you describe the relationship between Rocky and Tayo? What do the cattle represent for Josiah? How are the Mexican cattle different from other cattle? How would you describe Night Swan? What do you make of the section on 109–13? How would you describe Betonie? How does Silko depict "whiteness"?

F *Read*: Silko: *Ceremony* (138–70).

Think About: What's the importance of Betonie's ceremony for Tayo? What does Tayo learn?

Week 13 T *Read*: Silko: *Ceremony* (170–266).

Think About: How would you describe Ts'eh? What role does she play in Tayo's ceremony? How does Tayo change during the course of the novel? In what ways does the novel itself function as a healing ceremony? Paul Lauter described *Ceremony* as a text that's "an actor for change." Does this seem like an applicable description? Why/not? Why do you suppose *Ceremony* has, in many ways, become a literary classic?

F *Read*: Excerpts from *Boricuas*. Specific page numbers to be announced.

Final projects.

Week 14 T *Read*: Excerpts from *Boricuas*. Specific page numbers to be announced.

Final projects.

F No class.

Week 15 T *Read*: Excerpts from *Boricuas*. Specific page numbers to be announced.

Final projects.

F   *Read*: Excerpts from *Boricuas*. Specific page numbers to
be announced.

Final projects.

★   ★   ★

U.S./American Literature I: Precolonial
through Early Federal Period

Dr. AnaLouise Keating
English 370

You cannot spill a drop of American blood without spilling the
blood of the whole world. On this Western Hemisphere all tribes &
people are forming into one federated whole.

Herman Melville

Required Texts

Heath Anthology of American Literature, Vol. I.

Handout (HO)

Excerpt from *Genesis*; "Anne Hutchinson's Examination at the
Court . . . November 1637;" Gloria Anzaldúa: excerpt from *Borderlands/
La Frontera*.

Course Description

This course explores some of the earliest American literatures. We begin
with indigenous oral-based literatures; move on to survey the literatures
of discovery, conquest, and exploration; and conclude with an examination
of the literatures of settlement, colonization, and nation-building. We
focus on the variety of early American literatures and explore the relationship between these texts and the development of the U.S./American
literary canon.

Objectives

To become acquainted with and appreciate a wide variety of oral- and
written-based literary texts; to explore the ways class, culture, gender,

region, religion, and other differences might influence writing and shape identities; to reexamine our conceptions of "American literature" and literary merit; to develop powers of critical reading and thinking through in-class discussions and writing assignments; to become more familiar with the Internet; to develop cohesiveness as a classroom community; and to have fun.

As You Read

For each assignment, I've included questions and topics that we'll be discussing in class. (Marked *Think About.*) Please think about these issues as you read, and take notes! Here are some additional questions you might ask yourself as you read: Do you like the material? Why or why not? What don't you understand? Does it remind you of any other readings? How? What is the writer's view of human beings? of nature? of the divine? of "America"? How do the various assignments challenge previous conceptions of literature? Do they have "literary merit" and seem worthy of study in a literature class? Why or why not? Read each poem several times; look up words you don't know.

Because the course is structured both thematically and chronologically, it's your responsibility to be familiar with each writer's era. Although I have not listed the introductory material for each writer in the syllabus, I expect you to read it and will hold you accountable for the material on quizzes. But please—don't read this material until *after* you've read the assignment. It offers information that could be useful as you think about the writings, but sometimes it's overly opinionated. *All readings must be completed by the date listed in the syllabus.* Always bring the *Heath Anthology* to class.

Discussion Group

I have set up a listserv discussion group specifically for this course. This discussion format will allow us to continue our conversations outside of class. If you have questions about the reading, additional comments triggered by class discussions, or want to make sure we discuss a specific issue or text during class, write to the listserv. At times, I might suggest a question for discussion or specific readings that I want to make sure we talk about in class. I will also use the list for announcements, reminders, unexpected changes in readings, etc. Everyone must post at least one message to the listserv, and they must do so before the fourth week of class. Please make sure to include your name on all postings.

Discussion is an important part of the course, and it's vital that you attend all classes. Consequently, a significant portion of your grade is based on attendance. You should be aware that, if you miss class, you will lose attendance and quiz points. You are "allowed" five absences. I will deduct points for any additional absences. (See below.)

Grades

Grades will be awarded on a point system. You may achieve the required number of points for each letter grade in a variety of ways.

Attendance. Each class is worth 10 points. Points will be deducted for late arrivals and/or early departures. If you miss more than five classes, I will deduct points from your final grade (15 points for each additional class missed).

In-Class Essays. At least once a week, we'll begin class with a brief in-class essay based on the day's assigned readings. Generally, I will draw essay questions from the questions raised in the sections marked *Think About* and from questions and issues I suggest on the listserv. The essays will be closed book; however, you may use your own notes and your syllabus. Each essay will be worth up to 15 points, based on both content and form. (I reserve the right to give additional points for exceptionally insightful answers.) *Essays cannot be made up, so be on time!*

Discussion. Discussion is essential to the learning experience in this course. Because literature is always open to multiple interpretations often based at least partially on each reader's experiences and beliefs, it's important that we all respect each other's values, interpretations, and views. Personal opinions play a role in class dialogues but do not substitute for a sustained effort to understand and interpret all course material and to use textual evidence to support your interpretations. (Discussion is worth up to 200 points.)

Listserv. Each time you send a comment, question, or response to the listserv, you will receive at least 5 points (unless you simply respond to a previous comment by saying something like "I agree"). Up to 45 additional points might be awarded based on the amount of time and thinking reflected.

Interaction Papers. Typed, one- to two-page, single-spaced papers exploring one of the questions in the section marked *Think About*. As the word "interaction" suggests, you should actively respond to the readings assigned for each topic. In other words, don't just summarize the material; instead, use your own ideas, perspectives, and opinions as you interpret the texts. Generally, papers will be scored on a 30-point system [30–29: truly excellent; 28–26: very good; 25–22: good; 21–19: average; and so on.] Scores will reflect both content and form. Interaction papers must be turned in on the date of the assigned reading(s). For additional guidelines see the last page of this syllabus.

Final Exam. Cumulative take-home essay. Worth up to 100 points. Students scoring over 1100 points will be exempted from the final. (Talk about incentives!)

Grade:	A	B	C	D	F
Points:	837+	836–744	743–657	656–558	557 or below

*Schedule of Class Meetings**

Week 1 M Introduction: Definitions, course expectations, a brief history of "American" literature as a discipline.

First Americans

Th Lecture on Native American Worldviews.

"Talk Concerning the First Beginning;" excerpt from *Genesis* (HO).

Think About: Compare/contrast these two creation stories, focusing on what the similarities and differences might indicate about Native American and Christian worldviews. In the teacher's manual to this section, the editor mentions the "aesthetic frustration" U.S. students might feel while reading early Native narratives: In what ways does "Talk" violate western literary conventions (i.e., the types of literature you've read in other lit courses)?

Week 2 M "Changing Woman & the Hero Twins"

Think About: What does today's creation story indicate about the Navajo worldview? What similarities and differences do you see between this story, *Genesis*, and "Talk Concerning the First Beginning"?

Th "Wohpe & the Gift of the Pipe;" "The Origin of Stories;" "Raven Makes a Girl Sick."

---

* The following is a *tentative* schedule: I might change readings, due dates, or assignments, so please call or e-mail me if you miss class.

Think About: What might today's readings indicate about Native worldviews? What similarities and differences do you find between these readings and the previous readings?

Week 3 M "Iroquois or Confederacy of the Five Nations;" A Selection of Poems.

Think About: How does this story account for the Iroquois' origin? What values does this story praise? What values does it censure? Do you find any recurring themes in today's poems?

Th "The Singer's Art;" "Two Songs;" "Like Flowers . . . ;" "Song;" "Moved;" "Formula to Cause Death."

Think About: How do these poems depict nature? What similarities and differences do you find between these poems and the creation accounts we've examined?

Earliest Contacts

Week 4 M Columbus: From *Journal of the First Voyage to America, 1492–1493*; Handsome Lake: "How America Was Discovered."

Think About: How do today's readings represent "America?" In what ways does today's excerpt from Columbus change your view of Columbus? How would you describe the writing styles in today's readings?

Th Yuchi Tale: "Creation of the Whites;" Cabeza de Baca: from *Relation of Alvar Nuñez Cabeza de Baca*.

Think About: How does the Yuchi tale depict the interaction between Natives and the colonizers? What similarities and differences do you find between this tale and Handsome Lake's (182–84)? In what ways does de Baca's text function as "hagiography [the life of a saint], captivity narrative, & immigrant tale" (128). How does de Baca depict the Indians, nature, and God? What similarities and differences do you find between his views and Columbus's?

Week 5 M Gaspar Pérez de Villagrá: Canto I of *The History of New Mexico*; Fray Marcos de Niza: from *A Relation of the Reverend Father Fray Marcos de Niza* . . .

Think About: How do de Villagrá and de Niza view America? What similarities and differences do you find between their views, Columbus's, and Cabeza de Baca's? How do you account for similarities and differences?

Th Don Antonio de Otermín: "Letter on the Pueblo Revolt of 1680;" Hopi Version: "The Coming of the Spanish and the Pueblo Revolt."

Think About: Compare/contrast these two versions of the Pueblo Revolt. How does perspective influence each version? How are these representations of Indians and colonists similar to previous representations we've read for this course? How are they different?

Separatists, Puritans, and Others

Week 6 M William Bradford: *Of Plymouth Plantation*, chapter XIX; Thomas Morton: from *New England Canaan*.

Think About: Compare/contrast these two accounts of Merrymount. How does each writer's position influence his perspective and the way he narrates events? Do you find one account more reliable than the other? If so, why? If not, why not?

Th John Winthrop: from *A Modell of Christian Charity*; from *The Journal of John Winthrop*; "Anne Hutchinson's Examination at the Court . . . November 1637" (HO).

Think About: What does Winthrop mean when he describes his community as "a Citty [sic] upon a Hill" (233)? What does Winthrop's *Journal* indicate about the Puritan worldview? What does the court transcript (handout) tell us about Anne Hutchinson—both her personality and her beliefs? What does it tell us about Winthrop's personality and beliefs? Why do you imagine Hutchinson represented such a threat to Winthrop and many other Puritans?

Week 7 M Roger Williams: *A Key into the Language of America*; *New England Primer*.

Think About: How does Williams depict the Indians? What similarities and differences do you find between his depiction and those of other writers we've examined? How is God described in the *Primer*? How does the *Primer* depict human nature? What recurring themes do you find in the *Primer*?

Th Ann Bradstreet: "The Prologue;" "The Author to Her Book;" Edward Taylor: "Upon a Spider Catching a Fly;" "Huswifery;" "The Ebb & Flow;" "Prologue."

Think About: How do Bradstreet and Taylor view their poetry? How does Taylor represent his relationship with God?

Week 8 M Mary Rowlandson: Excerpts from *Narrative of the Captivity* . . .

Think About: How does Rowlandson depict her relationship between human beings and the divine? Do you find any similarities between her captivity narrative and Cabeza de Baca's? Do you notice any change in Rowlandson's attitude toward her captors? What was Rowlandson's purpose in writing her narrative? What impact do you suppose she might have wanted to make on her readers? How does her writing lend itself to a greater understanding of women's roles in Puritan life? How does being a woman affect her point of view? How might it affect the ways readers interpret her story?

Th Cotton Mather: "The Devil Attacks . . . ;" Lucy Terry: "Bars Fight."

Think About: How does Mather depict the witchcraft trials? Does he try to justify them? If so, how? What does he mean by "spectral evidence"? How does Terry depict the Indians?

Week 9 M Sarah Kemble Knight: *Journal.*

Think About: What does Kemble's *Journal* teach us about life in colonial America? How would you describe Kemble? In what ways is she similar to and different from other Puritan writers?

Th Elizabeth Ashbridge: From *Some Account of the . . . Life*; John Woolman: *Journal.*

Think About: Compare/contrast the ways these writers represent themselves, their relationships with the divine, and their spiritual discoveries/growth. How do their views differ from those of the Puritans? How are they similar?

Enlightenment Ideals

Week 10 M J. Hector St. John de Crevecoeur: From "What Is an American?;" Milcah Marth Moore: "The Female Patriots."

Think About: How does de Crevecoeur describe Americans? What's missing from his definition? What similarities and differences do you find between Crevecoeur's definition of an "American" and contemporary definitions? What type of rebellion does Moore advocate? What role do women play in this rebellion?

Th Jonathan Edwards: *Personal Narrative*.

Think About: Compare/contrast Edwards's worldview with those of earlier Puritans. How does Edwards define success? What role(s) does God play in his narrative? How does he view himself?

Week 11 M Benjamin Franklin: Excerpt from Poor Richard's *Almanacks*; excerpt from *Autobiography*.

Think About: What types of behavior are valued in the *Almanacks*? What similarities do you find between Franklin's narrative and others we've read for this course? (Topics to consider: individualism, religion, nature, success). How do you account for these similarities and differences? How does Franklin's *Autobiography* contribute to the myth of rugged individualism?

Th John and Abigail Adams: "Letters;" Thomas Paine: From *Common Sense*; Philip Freneau: "A Political Litany;" "On the Universality and Other Attributes of the God of Nature;" "Indian Burying Ground."

Think About: Based on these letters, how would you describe John and Abigail Adams (their values, their personalities, and the issues that concerned them)? How would you describe their relationship? How does Paine describe "American Affairs"? How does he describe the relationship between Great Britian and the American colonies? How does he justify his arguments? How does Freneau view America? How does he describe God?

Week 12 M Thomas Jefferson: from *Notes on the State of Virginia*; Prince Hall: "To the Honorable Council and House of Representatives . . ."

Think About: How does Jefferson view the indigenous American peoples? What similarities and differences do you find between his view and Freneau's? How does Jefferson view slavery? How does he justify his views about slavery and Native Americans? How might Hall have responded to Jefferson?

Th "Poems published anonymously;" Murray: "On the Equality of the Sexes."

Think About: What strategies do today's writers adopt in their arguments for changes in the relationships between men and women? Why do you suppose they employed these strategies?

Week 13 M Phillis Wheatly: "To the University of Cambridge;" "On Being Brought from Africa to America;" "Letter to Samson Occom;" Jupiter Hammon: "An ADDRESS to Miss Phillis Wheatly . . . ;" Samson Occam: "A Short Narrative of My Life."

Think About: How do Wheatly, Hammon, and Occam interact with each other? Do you find their perspectives similar? What do they have in common with the Puritans? How are they different? How might they have responded to Jefferson?

Th No class (Thanksgiving Holiday).

Narratives, Stories, and Social Critique

Week 14 M Sarah Morton: "The African Chief;" Olaudah Equiano: From *The Interesting Narrative* . . .

Think About: What similarities and differences do you find between Equinao's self-presentation and those in other first-person narratives (Rowlandson, Franklin, Edwards, etc.) we've read in this course? How do you account for these similarities and differences? How does Equiano describe African cultures? How does he view non-Africans? What impact does he want to have on his readers? Compare/contrast Equinao's representation of African culture and slavery with Morton's. What similarities and differences do you find?

Th Charles Brockden Brown: "Somnambulism."

Think About: How is "Somnambulism" different from other first-person narratives we've read in this course? How do you account for the differences? How reliable is the narrator? (That is, do you believe he gives an objective, unbiased accounts of events?) What, if anything, might influence his perceptions?

Week 15 M Bridget Fletcher: "The Greatest Dignity of a Woman;" "The Duty of Man & Wife;" Susanna Haswell Rowson: From *Charlotte Temple*.

Think About: How does *Charlotte Temple* function as social critique? (In other words, in what ways does Rowson seem to be analyzing and perhaps criticizing eighteenth-century society?) Why do you suppose this novel was so popular? What similarities and differences do you find between Fletcher's and Rowson's perspectives on marriage and women?

Th Mercy Otis Warren: "To a Young Woman;" Hannah Webster Foster: From *The Coquette*.

Think About: How does *The Coquette* function as social critique? Compare/contrast this novel with *Charlotte Temple*. Did you find its epistolary form difficult to follow? Why/not? What might Warren's poem reveal about her values? Compare/contrast these values with those in *The Coquette*.

Week 16 M "Virgin of Guadalupe;" Hispanic Southwest Tales: "La comadre sebastiana;" "El obispo;" "El indito de las cien vacas;" "Llorana;" Anzaldúa: Excerpt from *Borderlands/ La Frontera* (HO).

Think About: How do these stories "entertain" their listeners/readers into ethical behavior? What are the ethical behaviors the stories advocate? How does Anzaldúa revise the traditional stories? What purposes do her revisions serve?

Th Washington Irving: "Rip Van Winkle."

Think About: What types of behavior does Irving's story reward? What types of behavior does it censure? How does Rip's town change in the course of this story? How do you account for these changes? What points might Irving be trying to make?

★   ★   ★

Literature of the Southwest

Dr. AnaLouise Keating
English 413/593

Place is the source of who you are in terms of your identity, the language that you are born into and that you come to use.

Simon Ortiz

Course Description

We will examine a single area—the U.S. Southwest—exploring the dialogues that occur between people and the land; women and men; "Americans" and so-called others; and among Natives, Spanish, Mexican Americans, and Anglos. I will define "Southwest" narrowly, focusing only on Texas, New Mexico, and Arizona. This definition has everything to do with time restrictions and our own location in eastern New Mexico.

Texts, Required of All Students

Rudolfo Anaya: *Bless Me, Ultima*; Gloria Anzaldúa: *Borderlands/La Frontera: The New Mestiza*; Willa Cather: *Death Comes for the Archbishop*; Ana Castillo: *So Far From God*; David King Dunaway, ed.: *Writing the Southwest* (WS); Joy Harjo: *Secrets from the Center of the World*; Barbara Kingsolver: *The Bean Tree*; Nora Naranjo-Morse: *Mud Woman: Poems from the Clay*; Tomás Rivera: *y no se lo tragó la tierra/And the Earth Did Not Part*; Leslie Marmon Silko: *Ceremony*; Luci Tapahonso: *Blue Horses Rush In: Poems and Stories*; Frank Waters: *The Man Who Killed the Deer*.

Texts, Required of Graduate Students, Recommended for Undergraduates

The Portable Western Reader (Viking Portable Library), ed. William Kittredge.

Handouts

Mabel Major: "Introduction" to *Southwestern Heritage*; Rudolfo Anaya: "An American Chicano in King Arthur's Court."

Additional Handouts for Grad Students

Stuart Cochran: "The Ethnic Implications of Stories, Spirits, and the Land in Native American Pueblo and Aztlán Writing."

As You Read

For each assignment, I've included questions that we'll be discussing in class (marked *Think About*). Please take the time to consider these issues. Here are some additional questions you might ask yourself as you read: Do you like the material? Why or why not? What don't you understand? Does it remind you of any other readings? How? How does each writer depict the Southwest and the "spirit of the place"? How might gender, sexuality, color, class, and historical period shape each writer's perspectives? *All readings must be completed by the date listed in the syllabus.*

E-Mail Discussion Group

I have set up a listserv discussion group specifically for this course. This discussion format will allow us to continue our conversations outside of class. If you have questions about the reading, or additional comments triggered by class discussions, send them to the listserv. At times, I might suggest a question for discussion. I will also use the list for announcements, reminders, unexpected changes in readings, etc. Throughout the semester, graduate students will be posting brief discussions of their assigned readings.

Attendance Policy

Because this class meets only once a week, it's vital that you attend all classes. A significant portion of your grade is based on attendance. You should be aware that if you miss class you will lose attendance and quiz points.

Grades for Undergraduates

Grades will be awarded on a point system. You can achieve the required number of points for each letter grade in a variety of ways.

Attendance. Each class is worth 25 points. (Points will be deducted for late arrivals and early departures. The specific number of points deducted will depend on how late you arrive and how early you leave.)

Quizzes. There will be a series of unannounced quizzes, based on the evening's assigned readings. Generally, I will draw quiz questions from the questions raised in the sections marked *Think About* and from questions I suggest on the listserv. Each quiz will be worth at least 10 points. Quizzes cannot be made up.

Listserv. Each time you send a comment, question, or response to the listserv, you will receive at least 5 points. (Additional points might be awarded based on the amount of time and thinking reflected. You won't receive points if you simply respond to a previous comment by saying something like "I agree.")

Optional Interaction Papers. Typed, one- to two-page, single-spaced papers exploring one or more of the questions in the section marked *Think About.* As the word "interaction" suggests, you should actively respond to the readings assigned for each topic. In other words, don't just summarize the material; instead, use your own ideas, perspectives, and opinions as you interpret the texts. Generally, papers will be scored on a 25-point system [25–20: truly excellent; 20–17: very good; 15: good; and so on.] All scores will reflect both content and form. Papers must be turned in on the date of the assigned reading(s). For additional guidelines see the last page of this syllabus.

Final Exam. Take-home cumulative essay. Worth up to 75 points.

Grade:	A	B	C	D	F
Points:	760+	759–676	675–592	591–507	506 or below

Grades for Graduate Students

Grades will be awarded on a point system similar to that for undergraduates:

Attendance. See above.

Quizzes. See above.

Listserv. See above.

Readings and Class Reports. You will meet with me before class begins on 1/29 to determine additional readings from *The Portable Western Reader.* Throughout the semester, you will present brief reports, both orally and on the listserv, over these readings. Worth up to 25 points each.

Term Paper. Rather than completing optional interaction papers, you will complete a term paper (15 to 20 pages, typed) over some aspect of southwestern writing. (We will discuss paper requirements and topics in conference.) A research proposal is due 2/5; an annotated bibliography and progress report #1 is due 3/5; an outline is due 3/19; a rough draft is due 4/9; progress report #2 is due 4/30; and the final paper is due 5/14. This paper will be worth up to 500 points.

Final Exam. See above.

Grade:	A	B	C	D	F
Points:	1345+	1344–1196	1195–1046	1045–897	896 or below

*Schedule of Class Meetings**

Week 1 Introduction: Southwest of what? Defining the Southwest, Inventing a Region. The politics of canon formation and regional writing. Frontier, Wilderness, and the "Spirit of Place."

Week 2 *Read*: Mabel Major: "Introduction" to *Southwestern Heritage* (HO); Rudolfo Anaya: "Foreword: The Spirit of the Place" (WS ix–xvi); Paula Gunn Allen: "Preface" (WS xvii–xxiv); David Dunaway: "Introduction" (WS xxv–xlvii); Leslie Marmon Silko: *Ceremony* (Pages to be announced in class and on the listserv.) 1–102.

Think About: What does Anaya mean by the "spirit of the place"? What does Allen mean by "geospiritual" (WS xx)? How does Dunaway describe "a sense of place"? What type of "sense" or "spirit(s)" of place does Silko create in *Ceremony*? How would you describe Tayo at the outset of this novel? What connections do you see between the poetry and prose sections?

Week 3 *Read*: Leslie Marmon Silko: Finish reading *Ceremony* for tonight's class.

Think About: According to Allen, "[a] truly Southwestern work almost inevitably combines the ancient, the medieval, and the contemporary in ways that yield maximal meaning comprehensible within several contexts" (WS xxiii). Do you

---

* The following is a *tentative* schedule. I might change readings, due dates, or assignments, so please call me if you miss class.

see this combination in *Ceremony*? If so, how? How does Tayo change during the course of the novel? What roles do women play in his transformation? In what ways does the novel itself function as a healing ceremony?

Week 4 *Read*: Rudolfo Anaya: *Bless Me, Ultima*; "Rudolfo Anaya" (WS 16–30); Anaya: "An American Chicano in King Arthur's Court" (HO).

Grads: Stuart Cochran: "The Ethnic Implications of Stories, Spirits, and the Land in Native American Pueblo and Aztlán Writing" (HO).

Think About: How does Anaya depict the Southwest? How does his novel reflect and embody the spirit(s) of the place he discusses in his forward to *Writing the Southwest*? What roles do women play in Anaya's narrative? According to Ramón Saldívar, Ultima is used to support patriarchal structures. Do you agree? Why/not? Compare/contrast Anaya's uses of myth to Silko's. Do you find one more effective than the other? Why?

Week 5 *Read*: Nora Naranjo-Morse: *Mud Woman: Poems from the Clay* (Pages to be announced in class and on the listserv.) "Frank Waters" (WS 220–35); Frank Waters: *The Man Who Killed the Deer* (pages to be announced in class and on the listserv).

Think About: How does Noranjo-Morse describe her relationship with clay? How would you describe the relationship between her poetry and pottery? In what ways—if any—does her pottery enhance your understanding of her words, and in what ways—if any—do her words enhance your understanding of her pottery? How does Perlene challenge stereotypical views of Indian women? Does Waters stereotype women? If so, how? Compare/contrast Waters's depiction of Indians and "Hispanics" with Silko's and Anaya's. How do you account for similarities and differences? According to Dunaway, "[t]he symbolic representation of the natural world in myths is at the center of Waters's work" (WS 222). What forms do these representations take in *The Man Who Killed the Deer*?

Week 6 *Read*: Nora Naranjo-Morse: *Mud Woman: Poems from the Clay*; Frank Waters: *The Man Who Killed the Deer* (finish reading both books for tonight's class).

Think About: According to Anaya, Waters, more than many southwestern writers, created a "sense of synthesis" between Chicanos, Anglos, and Indians (WS 26). In what ways do you see this synthesis in Waters's novel? How does Waters depict Indians' view of the land? How is this depiction similar to and different from Silko's? Anaya also claims that "[t]he newer migrations into the Southwest brought not only different languages and religions, but a rugged individualism which stood in opposition to that [Native] communal spirit" (WS xiv). Do you see this conflict between individualism and communal sharing in Waters's novel? If so, how? How would you describe Waters's perspective on this conflict?

Week 7   *Read*: Willa Cather: *Death Comes for the Archbishop* (pages to be announced in class and on the listserv).

Think About: Erlinda Gonzales-Berry reads Cather's novel "as cultural artifact . . . [which] articulate[s] a selectivity atuned with the needs of the dominant power structure." According to Gonzales-Berry, "Perhaps no other Southwestern novel exemplifies this [imperialist] process as egregiously as Willa Cather's *Death Comes for the Archbishop*. In her portrayal of Father Martínez, for example, she inscribes in the creative domain the attitudes toward the native clergy disseminated by the texts of Gregg and Davis. Cloaking New Mexican priests in habits of licentiousness and immorality, the texts of the latter facilitated the Americanization of the Mexican Catholic church and its people. Consequently, whatever Cather's intent, her novel must be viewed as part of a larger discourse of domination" (5). How do you see this "discourse of domination" at work in Cather's novel? According to Anaya, "[t]he newer migrations into the Southwest brought not only different languages and religions, but a rugged individualism which stood in opposition to that [Native] communal spirit" (WS xiv). Do you see this conflict between individualism and communal sharing in Cather's novel? If so, how? How would you describe Cather's perspective on this conflict?

Week 8   *Read*: Luci Tapahonso *Blue Horses Rush In: Poems and Stories*; "Luci Tapahonso" (WS 198–219); Finish reading Cather's *Death Comes for the Archbishop*.

Think About: What similarities and differences do you see between Noranjo-Morse and Tapahonso? How does

Tapahanso represent a "double life" (WS 200–01)? How does she bring the two halves together? According to Dunaway, Tapahonso depicts a "spiritually alive landscape" and an "intimate physical connection" to the land (WS 204). Find specific examples in Tapahonso's poetry and prose where you see this view of nature and the land. Tapahonso claims that, in her poems and stories she "attempt[s] to create and convey the setting for the oral text" (*Blue Horses* xiv). Find and mark places where she seems to do this. Which poems and stories struck you as especially compelling? Why? How does Tapahonso's use of Navajo affect your reactions to her work? What is the significance of Cather's title? How does it shape your interpretation of the text?

Week 9 *Read*: Tomás Rivera: *y no se lo tragó la tierra/And the Earth Did Not Part*.

Think About: What is the significance of the title? How does it shape your interpretation of the text? How would you describe the narrator? How does Rivera depict the Southwest? How is his depiction similar to and different from the other writers we've read for this course? What roles does gender play in this novel? Would you describe *y no se lo tragó la tierra/And the Earth Did Not Part* as a novel? a collection of short stories? both? neither? According to Teresa McKenna, "Tomás Rivera explores the psychological and political dilemmas of migrant life, the movement from place to place, and its effects on family, religious belief, and sense of self." Do you agree with this interpretation? Why/not? What does Rivera seem to say about migrant life?

Week 10 *Read*: Gloria Anzaldúa: *Borderlands/La Frontera: The New Mestiza*: "Preface" and chapters 1–4 (1–52); Poems in sections I (101–13), V (176–89), and VI (192–203).

Think About: Compare/contrast Anzaldúa's Texas with Rivera's. How do you account for similarities and differences? How does Anzaldúa describe the Borderlands? What similarities and differences do you find between Anzaldúa's description of the southwest and other descriptions we've read in this course? What roles do gender and sexuality play in *Borderlands/La Frontera*? How does Anzaldúa redefine conventional gender roles? Compare/contrast Anzaldúa's poetry with the other poetry we've read for this course. What similarities and differences do you find? What are your reactions to Anzaldúa's code-switching?

Week 11   *Read*: Anzaldúa: *Borderlands/La Frontera*, chapters 5–7 (53–91); Poems in sections II (115–28), III (139–52), IV (153–73).

Think About: How does Anzaldúa attempt to synthesize the personal with the political? Do you find her synthesis effective? What's the relationship between Anzaldúa's "Shadow-Beast" and Coatlicue? How does Anzaldúa use myth to (re)construct her identities? Compare/contrast her use of myth with that of other writers in this course. What does she mean by "mestiza consciousness"? Which writers in this course seem to enact this form of thinking? How does Anzaldúa's description of la facultad and the interface (38) go with the poem "Interface" (148–52)? What other connections do you find between the *Borderlands/La Frontera*'s prose and poetry sections?

Week 12   *Read*: Ana Castillo: *So Far from God* (1–169).

Think About: Does Castillo's New Mexico resemble your view of this "land of enchantment"? Why/not? What similarities and differences you find between Anzaldúa's depiction of the Southwest and of gender relations in *Borderlands/La Frontera* and Castillo's depiction of them? What elements of Castillo's novel strike you as especially unrealistic? What points do you think Castillo is trying to make through these events? How would you describe Castillo's writing style? How would you describe Sofi and her daughters? How does Castillo use these characters to define female identity? How would you describe the narrator's voice?

Week 13   *Read*: Ana Castillo: *So Far from God* (170–252).

Think About: See issues for 4/16. How does Castillo represent religion? What roles does spirituality/religion play in creating, strengthening, and/or redefining female identity? Compare/contrast her views on gender and religion with those of other writers we've read for this course. How does Castillo redefine conventional definitions of motherhood? What is the significance of Castillo's title? How does it shape your interpretation of the text?

Week 14   *Read*: "Barbara Kingsolver" (WS 93–107); Barbara Kingsolver: *The Bean Trees* (pages to be announced in class and on the listserv).

Think About: How does Kingsolver depict the Southwest? Compare/contrast her depiction with those of other writers for this course. How would you describe Taylor, Turtle, and Lou Ann? How does Kingsolver depict female identity and gender relations?

Week 15   *Read*: Barbara Kingsolver: Finish reading *The Bean Trees* for this evening's class; Joy Harjo: *Secrets from the Center of the World*; "Joy Harjo" (WS 46–61).

Think About: Is *The Bean Trees*, as one critic suggested, about "invasion" (WS 100)? Why/not? In what ways is Kingsolver's novel "political" (WS 101–02)? How do Taylor, Turtle, and Lou Ann change in the course of the novel? To what do you attribute their changes? What is the significance of Kingsolver's title? How does it shape your interpretation of the text? What does Harjo mean by "sacred space" (WS 46)? How does she create this space in *Secrets*? Harjo emphasizes the importance of transformation in her work. How do you see transformation in her poetry (in *Writing the Southwest*) and in *Secrets*? How does *Secrets* convey "a feeling of space and freedom" (WS 53)?

Week 16   Final Exam Due

We will finish discussing Harjo. Each student will briefly discuss his/her final exam.

A few of the many books we're not reading—but "should": Charles F. Lummis: *A New Mexico David* (1891); *The Land of Poco Tiempo* (1893); Mary Austin: *The Land of Little Rain* (1903); Nina Otero de Warren: *Old Spain in Our Southwest* (1936); Cleofas M. Jaramillo: *Shadows of the Past/Sombras del Pasado* (1941); Fabiola Cabeza de Baca: *We Fed Them Cactus* (1954); Edward Abbey: *Desert Solitaire* (1968); N. Scott Momaday: *House Made of Dawn* (1968); Tony Hillerman: *The*

Blessing Way (1970); John Nichols: *The Milagro Beanfield War* (1974); Sandra Cisneros: *Woman Hollering Creek* (1991); Denise Chávez: *Face of An Angel* (1994).

★ ★ ★

Feminist Epistemologies

Dr. AnaLouise Keating
WS 5363

The unleashed power of the atom has changed everything but our way of thinking; and so we are headed for an unparalleled catastrophe. . . . A new type of thinking is indispensable if [hu]mankind is to survive and develop further.

Albert Einstein

In trying to become "objective," Western culture made "objects" of things and people when it distanced itself from them, thereby losing "touch" with them. This dichotomy is the root of all violence. Not only was the brain split into two functions but so was reality. Thus people who inhabit both realities are forced to live in the interface between the two, forced to become adept at switching modes.

Gloria E. Anzaldúa

Feminist epistemology suggests that integrating women's contributions into the domain of science and knowledge will not constitute a mere adding of details; it will not merely widen the canvas but result in a shift of perspective enabling us to see a very different picture. The inclusion of women's perspectives will not merely amount to women participating in greater numbers in the existing practice of science and knowledge, but it will change the very nature of these activities and their self-understanding.

Uma Narayan

The greatest revolution of our time is in the way we see the world. The mechanistic paradigm underlying the Industrial Growth Society gives way to the realization that we belong to a living, self-organizing cosmos. General systems theory, emerging from the life sciences, brings fresh evidence to confirm ancient, indigenous teachings: the Earth is alive, mind is pervasive, all beings are our relations. This realization changes everything.

It changes our perceptions of who we are and what we need, and how we can trustfully act together for a decent, noble future.

Joanna Macy

Course Description

Employing feminist/womanist and indigenous critiques of Enlightenment-based epistemologies, this course explores dominant and subjugated knowledge systems. We examine issues such as the following: Is reason gendered and/or "raced"? In what ways, if any, do social, geographical, historical, and bodily location affect knowledge production and consumption? How are knowledge, perception, belief, action, and power interrelated? What "counts" as knowledge, as fact/fiction, as truth/story? What are the justifications for these different designations, and are such distinctions always useful? How do feminist/womanist/indigenous knowledges alter or subvert research materials and methodologies in conventional academic disciplines? Do these challenges affect how you think of research materials and methodologies in your own field(s)? What kinds of knowledge are transformative and for whom? Should knowledge, necessarily, have transformative potential—in other words, whose interests are/should be/could be served by diversifying what "counts" as knowledge and truth?

Course Goals/Student Learning Outcomes

Students who successfully complete this course will obtain an increased understanding of the following: feminist/womanist/indigenous epistemologies and interventions into dominant knowledge systems; the relationship between power and knowledge; and epistemological debates, variations, and controversies within feminist theory. Students will also be able to: (1) evaluate discipline-specific epistemological framework(s); (2) understand, evaluate, and appreciate a diverse range of representations or understandings of knowledge claims and knowledge formations; and (3) articulate a definition of feminist epistemology as an area of study within Women's Studies and relative to their own disciplinary field(s). Students will obtain an increasingly complex ability to think relationally and an increasingly nuanced understanding of the ways commonalities, similarities, and differences work together; improved writing skills; and enhanced critical thinking skills.

E-Mail Discussion Group

I've set up a listserv discussion group specifically for this course. If you have questions about the readings, or additional comments triggered by class discussions, post them here! At times, I might suggest a question for discussion. This discussion format will allow me to send announcements, reminders, unexpected changes in readings, etc.

Required Texts

Patricia Hill Collins: *Black Feminist Thought* (2nd edition)
Gloria E. Anzaldúa: *Borderlands/La Frontera: The New Mestiza** (2nd edition)
Anna Lee Walters: *Ghost Singer**
Barbara Neely: *Blanche on the Lam*
Linda Alcoff and Elizabeth Potter, eds.: *Feminist Epistemologies**
Andermahr, Lovell, Wolkowitz: *A Glossary of Feminist Theory*
Michelle Le Doeuff: *The Sex of Knowing*

Recommended Texts

Joanna Macy: *Mutual Causality in Buddhism and General Systems Theory: The Dharma of Natural Systems*
George Sefa Dei et al.: *Indigenous Knowledges in Global Contexts: Multiple Readings of Our World** (IK)

Course Packet (CP)

Mary Ballou: "Women and Spirit: Two Nonprofits in Psychology;" Elizabeth Grosz: "The In(ter)vention of Feminist Knowledges;" Owen J. Dwyer and John Paul Jones III: "White Socio-Spatial Epistemology;" Linda Tuhiwai Smith "Colonizing Knowledges;" Genevieve Lloyd: "Reason as Attainment" and "Reason as Progress;" Joanna Macy: "The Great Turning;" Mary E. Hess: "White Religious Educators and Unlearning Racism;" Caroline New: "Realism, Deconstruction, and the Feminist Standpoint;" Nancy Pineda-Madrid: "Notes toward a Chicanafeminist Epistemology;" Paula Gunn Allen: "The Sacred Hoop;" Marilyn Frye: "White Woman Feminist; 1983–1992."

---

* On reserve at the Blagg-Huey Library; *Mutual Casualty* is available via the library's electronic books net library.

Supplemental Course Packet (SCP)

James J. Scheurich and Michelle D. Young: "Coloring Epistemologies: Are Our Research Epistemologies Racially Biased?;" Inés Hernández-Ávila: "Mediations of the Spirit: Native American Religious Traditions and the Ethics of Representation;" Melissa J. Fiesta: "Solving Mysteries of Culture and Self;" Kathryn Shanley: "Time and Time-Again: Notes toward an Understanding of Radical Elements in American Indian Fiction;" Uma Narayan: "The Project of Feminist Epistemology: Perspectives from a Nonwestern Feminist;" Marilyn Frye: "On Being White . . . ;" Arleen Dallery: "The Politics of Writing (the) Body: *Ecriture Feminine*;" Aída Hurtado: "Theory in the Flesh: Toward an Endarkened Epistemology;" Cynthia B. Dillard: "The Substance of Things Hoped For, The Evidence of Things Not Seen: Examining an Endarkened Feminist Epistemology in Educational Research and Leadership;" Madigan, Johnson, and Linton: "The Language of Psychology: APA Style as Epistemology."

As You Read

WS 5363 is a graduate-level course, and I expect all students to follow graduate-level academic practices: (1) I expect you to complete all readings by the date listed in the syllabus; (2) I expect you to read the material thoughtfully and in an engaged manner (take notes, reflect on the material, etc.); (3) I expect you to read all endnotes and footnotes; (4) I expect you to read (not skim) *all* of the required readings—even those you find "boring" or difficult; (5) I expect you to reread those texts that you have previously read; (6) I expect you to seek out definitions for words and terminology you don't know: Start with *A Glossary of Feminist Theory*; if the words don't appear in the *Glossary*, try the following websites:

http://plato.stanford.edu/contents.html
http://www.theory.org.uk/
http://www.uoguelph.ca/culture/glossary.htm
http://social.chass.ncsu.edu/wyrick/debclass/gloss.htm
http://www.popcultures.com/

If the terms do not appear in any of these sites, e-mail our class listserv.

   This reading-intensive course takes a nonlinear, comparative, dialogic approach. Throughout the course we will be asking the questions listed above under "Course Description," so please think about them as you read. Please read all poems several times.

Grades

Grades will be awarded on a point system. *It is your responsibility to keep track of your number of points.* Please do not e-mail or call me at the end of the semester to find out your grade.

Attendance. Each class is worth 15 points. (Points will be deducted for late arrivals and early departures. The specific number of points deducted will depend on how late you arrive and/or how early you leave.) A significant portion of your grade is based on attendance. You are "allowed" two absences (although you will still lose attendance points for these absences). If you miss three or four classes, you will be unable to receive any grade higher than a B. If you miss more than four classes, you will be unable to receive any grade higher than a C.

Participation. Participation entails both engaged, respectful listening and discussion. Personal opinions play a role in class dialogues but do not substitute for a concerted effort to grasp the scholarly material in the course. Because it's important that we all respect each other's needs, values, and views, our conversations will be guided by open-minded listening to divergent opinions. As June Jordan states: *"Agreement is not the point. Mutual respect that can accommodate genuine disagreement is the civilized point of intellectual exchange, particularly in a political context."* For more on this topic, see the end of the syllabus. Discussion takes two forms: in-class dialogue and electronic dialogue via our class listserv. (Participation is worth up to 50 points.)

Discussion Questions. For our classes on Weeks 2, 3, 5, 6, 8, 9, 10, 11 you will design a discussion question (questions you have about the readings that you'd like us to explore in class). Questions can deal specifically with one of the assigned readings or they can be a bit broader and engage several readings. Your discussion questions will significantly guide our class conversations and serve at least three additional purposes: (1) They offer opportunities for students to reflect more deeply on the assigned readings and, through this reflection, to deepen their learning. (2) They enhance student accountability and give students more control over our time together. (3) They allow me to assess student interests and learning. *Question format instructions*: If you ask a question about a specific passage, please provide the quotation and page number(s); if you ask a question about an issue found on specific pages, please include the page numbers. (There is no need to provide full bibliographical information for these questions.) Mark with an asterisk (★) those questions that you really hope we'll have time to discuss. E-mail your questions to me in two formats: (1) pasted into the body of the e-mail and (2) saved in Wordperfect or Word and sent as an attachment. (Just send one e-mail; include both formats in a single e-mail.) Do not ask definition-based or other easily researched questions (for instance: "What does 'episteme' mean?" or "What is Gender Studies?"). Each week's discussion question is worth up to 5 points, based on both content (50%) and form (50%), and will be due on or before *Sundays at noon*. (I'll give you an extra 5 points if I receive your questions by *Friday at 6 p.m.*) Please do not wait until the last minute to send your question; I will not give credit for late questions even if their tardiness is due to technical difficulties.

Talking Notes Handout. For our classes on 4/17 and 4/21 you will be asked to read an essay, prepare a handout, and present five-minute report on it. For the Talking Notes on 4/21, you will need to bring copies for the class. (You may use the copier in the Women's Studies office to make your copies.) Your handout may be *no longer than two pages*. (If you want to include the maximum amount of material, your handout may be single-spaced in 11-inch Arial font.) At the top of the page, include your text's title, author, and full page numbers. Do not try to write a brief essay. Instead, please follow this format:

Summary: Succinctly summarize the text, focusing especially on the main argument. Your summary should be only two to five sentences. Don't just give details; interpret the text. Please be sure to explain why you did or did not find the argument persuasive.

Outline: Outline the essay structure (include page numbers for each section).

Quotations: Select one or two key quotations that best illustrate the text. Briefly explain why they illustrate the text.

Connections: List other course authors/readings which your selected text resembles. Briefly explain why you see this resemblance.

These handouts serve several purposes, including: (1) they allow us, as a class, to cover more material; (2) they enable students to work on critical thinking, reading, and writing skills. Each handout will be worth up to 20 points (based on both content and form). In order to receive credit for these assignments, you must be in class on the day they are due. You will select your texts in class on Week 2.

Disciplinary Epistemological Analysis. This assignment requires you to research your discipline's conventional/traditional epistemological framework. (If you are a graduate student in women's studies, I recommend that you select your undergraduate discipline. If you feel too alienated from your discipline, you may focus on your higher-education experience more generally and discuss traditional western education's conventions. If you are a graduate student in another discipline, you should analyze that discipline's epistemology.) You should be able to answer the following questions: How does the discipline define knowledge? What methodologies (logic? analysis? reason? previous authorities? intuition? etc.) do members of your discipline use and value most highly as they attempt to arrive at knowledge and truth? What types of information do they most value (facts, experience-based narratives, poetry, previous research, etc.)? Bring to class a five- to eight-page (typed, double-spaced) discussion summarizing what you have learned. In addition to this discussion, please also include a list of Works Consulted containing at least ten references to outside sources. To ensure success on this assignment, I recommend the following: (1) Start early and spend a substantial amount of time writing and revising your analysis. (2) Extensively research your discipline's (or, more generally, western education's) epistemology. (3) Do not focus on the discipline per se (i.e., don't provide a history or extensive definition of the discipline); rather, focus specifically on the discipline's epistemology. (4) Explore, in detail, the epistemological implications of your statements. (5) Do not *defend* your discipline; *analyze* its epistemology. Please see our Blackboard site for a sample student paper. (Worth up to 100 points, based on both content and form. Late papers will be marked down, 10 points per day.)

Final. Drawing on the assigned class readings and your Disciplinary Epistemological Analysis, you will write a typed, six- to eleven-page explorative essay describing your personal epistemological perspective as it intersects with and diverges from: (1) the western dominant-culture epistemological system; (2) your discipline's (or undergraduate education's) epistemological framework; and (3) feminist, womanist, and/or indigenous epistemologies. As you develop your essay, I recommend that you look over all of your in-class freewrites and the questions included in the "Course Description" (above). Your final should include your definitions/ discussions of knowledge, truth, intuition, reason, emotion, and other key epistemological terms (as indicated in the freewrites and class discussions). As part of this assignment, you must complete the Final Essay Freewrite (located on our Blackboard site). *Failure to turn in this freewrite with your final essay will result in a reduction of 25 points from your grade!* You will also present a brief oral report summarizing your epistemology. Worth up to 150 points, based on both content and form. (I reserve the right to award additional points for exceptional work.) Papers will be scored according to the criteria listed above. Late essays will be marked down, 10 points per day.

Grade:	A	B	C	D	F
Points:	562 or above	561–496	495–435	434–375	374 or below

Calculation of Points

Attendance	15×15	225 points
Discussion		50 points
Discussion Questions	8×5	40 points
Disciplinary Analysis		100 points
Talking Notes (4/17)		20 points
Talking Notes (4/24)		20 points
Final		150 points
Total points possible		605 points

Freewriting Activities

Throughout the semester we will (probably) engage in a series of freewriting exercises. During these activities, which will generally last for ten minutes, you should write for the entire time. (Keep the pen moving even if you feel that your logical mind has nothing to say!) Please save all of your freewriting exercises. They should be useful as you write your essay(s). Here are some of the things you'll write about during your first freewrite:

How would you define knowledge and truth? What roles do logic, reason, analysis, intuition, and imagination play in what you consider knowledge to be? Is knowledge created? Is truth created? Are there universal truths? If so, who decides what these truths are and how do they know these truths? Are there objective truths? What roles, if any, does our subjectivity (our individual consciousness/identity/ etc.) play in knowing? How is knowledge proved: through sense experience? through external validation by authorities (scientists, religious leaders, history, angels, etc.)? by appeal to one's own consciousness (an inner voice? a hunch? etc.). How does knowledge influence daily life—on individual and social levels?

*Schedule of Class Meetings**

Week 1 Introduction (syllabus, background information on western epistemologies, self-reflection exercise).

Week 2 *Read*: Alcoff and Potter: "Introduction" (FE 1–14); Grosz: "The In(ter)vention of Feminist Knowledges" (CP 1–13); Patricia Hill Collins: "Preface to 1st Edition;" "Preface to 2nd Edition;" "The Politics of Black Feminist Thought" (1–20); "Distinguishing Features of Black Feminist Thought" (21–43); Patricia Hill Collins: "Work, Family, and Black Women's Oppression" (45–68); "Mammies, Matriarchs, and Other Controlling Images" (69–96); "The Power of Self-Definition" (97–122); "The Sexual Politics of Black Womanhood" (123–48).

Recommended: Macy: "Preface" (xi–xviii); "Introduction" (1–3); "Considering Causality" (7–21).

Note: Before today's class please look through *Feminist Epistemologies* and select two essays that you'd like to use for your Talking Notes Handouts.

Week 3 *Read*: Collins: "Black Women's Love Relationships" (149–72); "Black Women and Motherhood" (173–200); "Rethinking Black Women's Activism" (201–25); "US Black Feminism in Transnational Context" (227–50); "Black Feminist Epistemology" (251–72); "Toward a Politics of Empowerment" (273–90); Uma Narayan: "The Project of Feminist Epistemology: Perspectives from a Nonwestern Feminist" (SCP 91–97); Marilyn Frye: "White Woman Feminist; 1983–1992" (CP 147–57).

Recommended: Caroline New: "Realism, Deconstruction, and the Feminist Standpoint" (CP 73–96); Harding: "Rethinking Standpoint Epistemology" (FE 49–82); Frye: "On Being White . . ." (SCP 106–15).

Week 4 *Read*: Neely: *Blanche on the Lam* (no discussion question due this week).

Week 5 *Read*: Genevieve Lloyd: "Reason as Attainment" (CP 14–23); Le Deouff: *The Sex of Knowing* (ix–68); Arleen

Dallery: "The Politics of Writing (the) Body: Ecriture Feminine" (SCP 98–105).

Recommended: Lloyd: "Reason as Progress" (CP 24–32).

Week 6　*Read*: Paula Gunn Allen: "The Sacred Hoop" (CP 165–74); Linda Tuhiwai Smith "Colonizing Knowledges" (CP 118–27); Inés Hernández-Ávila: "Mediations of the Spirit: Native American Religious Traditions and the Ethics of Representation" (SCP 14–37); Joanna Macy: "The Great Turning" (CP 33–44); Walters: *Ghost Singer* (xi–79).

Recommended: Dei, Hall, Rosenberg: "Preface" (IK xi–xvi) and "Introduction" (IK 3–17); Shiva: "Foreword" (IK vii–x); Macy: "The Buddhist Teaching of Dependent Co-Arising" (25–43).

Week 7　*Read*: Walters: *Ghost Singer* (finish reading the novel for today's class).

Note: No discussion question due this week.

Recommended: Melissa J. Fiesta: "Solving Mysteries of Culture and Self" (SCP 38–43); Kathryn Shanley: "Time and Time-Again: Notes toward an Understanding of Radical Elements in American Indian Fiction" (SCP 44–51); Wangoola: "Mpambo, the African Multiversity: A Philosophy to Rekindle the African Spirit" (IK 265–77).

Week 8　*Read*: Anzaldúa: *Borderlands/La Frontera: The New Mestiza*: "Preface" and chapters 1–4 (19–73); Poems in sections I (123–35), II (137–60), and III (161–74); Mary E. Hess: "White Religious Educators and Unlearning Racism" (CP 58–72).

Recommended: Macy: "General Systems Theory" (69–89).

Week 9　*Read*: Anzaldúa: *Borderlands/La Frontera*, chapters 5–7 (75–113); Poems in sections IV (175–95), V (197–211), and VI (213–25); Nancy Pineda-Madrid: "Notes toward a Chicanafeminist Epistemology" (CP 104–17).

Recommended: Macy: "Mutual Causality in General Systems Theory" (91–104).

Week 10　*Read*: Cynthia B. Dillard: "The Substance of Things Hoped For, The Evidence of Things Not Seen: Examining an Endarkened Feminist Epistemology in Educational Research and Leadership" (CP); Aída Hurtado: "Theory in the Flesh: Toward an Endarkened Epistemology" (CP); Owen J. Dwyer and John Paul Jones III: "White Socio-Spatial Epistemology" (CP 45–57); James J. Scheurich and Michelle D. Young: "Coloring Epistemologies: Are Our Research Epistemologies Racially Biased?" (SCP 1–13); Mary Ballou: "Women and Spirit: Two Nonprofits in Psychology" (CP 97–103).

Note: Please bring *Borderlands/La Frontera* to class.

Week 11   *Read*: Paula M.L. Moya: "Postmodernism, 'Realism,' and the Politics of Identity: Cherríe Moraga and Chicana Feminism" (CP 128–46); Madigan, Johnson, and Linton: "The Language of Psychology: APA Style as Epistemology" (SCP 148–56).

Week 12   *Due*: Disciplinary Epistemological Analysis. Please bring two copies.

Note: Please bring last week's readings to class. No discussion questions due this week.

Week 13   *Read*: Gloria E. Anzaldúa: "now let us shift . . . the path of conocimiento . . . inner work, public acts" (SCP 52–90); Le Deouff: *The Sex of Knowing* (143–220).

Due: Talking Notes analysis of Anzaldúa's essay (Worth up to 20 points).

Note: No discussion questions due this week.

Recommended: Macy: "Self as Process" (107–16); "Mutual Morality" (193–213).

Week 14   *Read*: One essay from *Feminist Epistemologies*, selected. Come prepared to give a 5-minute summary/analysis.
Due: Talking Notes handout based on your selected essay. (Please bring copies for the whole class.) (Worth up to 20 points.)

Note: No discussion questions due this week.

Week 15   *Due*: Final Paper. Platica.

Recommended: Joseph Couture: "Native Studies and the Academy" (IK 157–67); Roxana Ng: "Toward an

Embodied Pedagogy: Exploring Health and the Body through Chinese Medicine" (IK 168–83).

★ ★ ★

U.S. Women of Colors

Instructor: Dr. AnaLouise Keating
WS 5463–01; Tuesdays 2:30 to 5:20 p.m.; HDB 309

Racism is a gaze that insists upon the power to make others conform, to perform endlessly in the prison of prior expectation, circling repetitively back upon the expired utility of the entirely known. Our rescue, our deliverance perhaps, lies in the possibility of listening across that great divide, of being surprised by the Unknown, by the unknowable. Old habits of being give way, let us hope, to a gentler genealogy of Grace.

Patricia J. Williams (1997)

We take our stand on the solidarity of humanity, the oneness of life, and the unnaturalness and injustice of all special favoritisms, whether of sex, race, country, or condition. If one link of the chain be broken, the chain is broken. . . . The colored woman feels that woman's cause is one and universal; and that not till the image of God, whether in parian or ebony, is sacred and inviolable; not till race, color, sex, and condition are seen as the accidents, and not the substance of life; not till the universal title of humanity to life, liberty, and the pursuit of happiness is conceded to be inalienable to all; not till then is woman's lesson taught and woman's cause won—not the white woman's, nor the black woman's, nor the red woman's, but the cause of every man and of every woman who has writhed silently under a mighty wrong. Woman's wrongs are thus indissolubly linked with all undefended woe, and the acquirement of her "rights" will mean the final triumph of all right over might, the supremacy of the moral forces of reason, and justice, and love in the government of the nations of earth.

Anna Julia Cooper (1893)

Course Description

This reading-intensive seminar adopts a relational approach to explore U.S. women-of-colors histories, theories, cultures, consciousness, and lives from a variety of perspectives. Readings, discussions, and writing assignments will focus especially on feminist/womanist issues related to conflict, agency, survival, resistance, intervention, and transformation.

Course Objectives

Students who successfully complete this course will obtain an increased understanding of the following: (1) the intersectionality of identities, especially as this intersectionality impacts U.S. women of colors; (2) women-of-colors histories, contemporary issues, and interventions into "mainstream" feminist theory; (3) the history and contemporary status and dynamics of "race" and racism in U.S. culture, including women's movements feminist theory, and women's studies; (4) womanist/feminist theorizing; and (5) research skills enabling them to continue learning about U.S. women-of-colors issues. Students will also obtain: (1) an increased ability to think relationally; (2) an increasingly nuanced understanding of the ways commonalities, similarities, and differences work together; (3) improved writing skills; and (4) enhanced holistic-critical thinking skills.

Required Texts

This Bridge Called My Back: Writings by Radical Women of Color, ed. Cherríe
   L. Moraga and Gloria E. Anzaldúa (3rd printing; 2002)
this bridge we call home: radical visions for transformation,* ed. Anzaldúa and
   Keating
*Kindred** by Octavia Butler
Last Standing Woman by Winona LaDuke
*Pedagogies of Crossing: Meditations on Feminism, Sexual Politics, Memory,
   and the Sacred*: M. Jacqui Alexander
A People's History of the United States by Howard Zinn

Handout (HO)

Beth Roy: "For White People, on How to Listen When Race is the Subject;" Kimberlé Crenshaw: "1989. Demarginalizing the Intersection of Race and Sex: A Black Feminist Critique of Antidiscrimination Doctrine, Feminist Theory and Antiracist Politics;" Mariana Ortega: "Being Lovingly, Knowingly Ignorant: White Feminism and Women of Color."

---

* The books marked with an asterisk are also on reserve at the library. *This Bridge Called My Back* is out of print, but I have copies available for purchase.

Reading

I expect you to read carefully, thoughtfully, and reflectively, and I suggest that you underline, make notes in the margins of your texts, and keep a reading journal. Please read all poems several times. All readings must be completed by the date listed in the syllabus. Always bring the assigned text(s) to class.

Grades

Grades will be awarded on a point system. *It is your responsibility to keep track of your number of points.* Please do not e-mail or call me at the end of the semester to find out your grade.

Attendance. Each class is worth 15 points. (Points will be deducted for late arrivals and early departures. The specific number of points deducted will depend on how late you arrive and/or how early you leave.) A significant portion of your grade is based on attendance. You are "allowed" two absences (although you will still lose attendance points for these absences). If you miss three or four classes, you will be unable to receive any grade higher than a B. If you miss more than four classes, you will be unable to receive any grade higher than a C.

Participation. Participation entails both engaged, respectful listening and informed discussion. Personal opinions play a role in class dialogues but do not substitute for sustained effort to understand and interpret all course material. Because it's important that we respect each other's needs, values, and views, our conversations will be guided by open-minded listening to divergent opinions. As June Jordan states: "*Agreement is not the point. Mutual respect that can accommodate genuine disagreement is the civilized point of intellectual exchange, particularly in a political context.*" (For more on this topic, see pages 9 to 10 in the syllabus.) Discussion takes two forms: in-class dialogue and electronic dialogue on Blackboard. I have set up, on the Discussion Board, a Forum called *EXPRESS YOURSELF!* I hope that you will use this forum to share your ideas, thoughts, reflections, and so on about the issues raised in this course. Occasionally I might suggest topics for discussion. (Participation is worth up to 50 points.)

Discussion Questions. For our classes on Weeks 2, 3, 5, 7, and 11, you will design one discussion question (a question you have about the week's readings that you'd like us to explore in class). The question can deal specifically with one of the assigned readings, or it can be a bit broader and engage several readings. Discussion questions will significantly guide our class conversations and serve at least three additional purposes: (1) They offer opportunities for students to reflect more deeply on the assigned readings and, through this reflection, to deepen their learning. (2) They enhance student accountability and give students more control over our time together. (3) They allow me to assess student interests and learning. *Question format instructions*: If you ask a question about a specific passage, please provide the quotation and page number(s); if you ask a question about an issue found on specific pages, please include the page numbers. (There is no need to provide full bibliographical information for these questions.) Mark with an asterisk (★) questions that you really hope we'll have time to discuss. E-mail your question to me in two formats: (1) pasted into the body of the email and (2) saved in Wordperfect or Word and sent as an attachment. (Just send one e-mail; include both formats in a single e-mail.) Here are two sample discussion questions:

1. In "Uses of the Erotic," Lorde writes: "The erotic is a resource within each of us that lies in a deeply female and spiritual plane, firmly rooted in the power of our unexpressed or unrecognized feeling" (53). How did you define the "us" in her sentence? Does it include all human beings and, if so, what implications does the fact that the erotic is located in a "deeply *female*" plane say about Lorde's understanding of gender?

2. In "Transformation of Silence Into Language and Action" Lorde challenges us to break silence and speak out, but in "Imagining Differently: The Politics of Listening in a Feminist Classroom" Cervenak et al. suggest that we need to listen, rather than speak out. (Or am I misinterpreting these pieces?) What are we to do: speak out? listen? both? neither? I'm interested in hearing what my classmates think about this apparent contradiction especially as it might impact our class discussions and, more generally, feminist theorizing.

Do not ask definition-based or other easily researched questions (for instance: "What does the word 'hegemonic' mean?" or "What is the difference between feminism and womanism?" or "When did the Trail of Tears occur, and how many people did it affect?"). Each week's discussion question is worth up to 5 points, based on both content (50%) and form (50%), and will be due on or before *Mondays at 9 a.m.* (I'll give you an extra 5 points if I receive your questions by *Saturday at 9 p.m.*) Please do not wait until the last minute to send your question; I will not give credit for late questions even if their tardiness is due to technical difficulties.

Talking Notes Handout/Discussion. For our classes on Weeks 6, 8, 9, 10, and 14, you will be asked to read an essay of your choice from specified sections in one of the course texts (as indicated on the schedule listed later in this syllabus), and prepare a handout. You will need to post your handout on Blackboard and also give a five-minute report on it. You must post your Talking Notes on Blackboard before class begins, and you will need to bring copies for the class. (You may use the copier in the Women's Studies office to make your copies.) For Talking Notes # 1–4: Your handout may be no longer than two pages; for Talking Notes #5: your handout may be no longer than four pages. (If you want to include the maximum amount of material, your handout may be single-spaced in 11-inch Arial font.) At the top of the page, include your text's title, author, and full page numbers for the essay. For example:

Laura Harris: "Notes from a Welfare Queen in the Ivory Tower" (*this bridge* 372–81).

Do not try to write a brief essay. Instead, please follow this format and use the indicated headings:

Summary: Summarize the text, focusing especially on the main argument. Don't just give details; interpret the text. Please be sure to explain why you did or did not find the argument persuasive.
Outline: Outline the essay structure (include page numbers). Your outline should summarize the section.
Themes: List and briefly describe the essay's main themes. Be sure to also discuss how the text fits (or does not fit) with the specific section in which it is located.
Quotations: Select one or two key quotations that best illustrate the text. Briefly explain why they illustrate the text.
Connections: List other course authors/readings that your selected text resembles. For each author/reading, briefly explain why you see this resemblance.

These handouts serve several purposes, including: (1) they allow us, as a class, to cover more material; (2) they enable students to work on critical thinking, reading, and writing skills. Each handout will be worth up to 15 points (based on both content and form). In order to receive full credit for these assignments, you must be in class on the day they are due. You will select your texts for Talking Notes #1–4 in class on 9/12, and you will select your chapter for Talking Notes #5 on 11/14.

(Talking Notes #1–4 are worth up to 15 points, based on both content and form; Talking Notes #5 is worth up to 25 points, based on both content and form.)

Intersectionality Essay. A typed, ten- to fifteen-page, double-spaced essay, tracing a theme or topic through selected course texts, focusing especially on issues of intersectionality. (For information on intersectionality, see Crenshaw's article and http://depts.washington.edu/ctp/Intersectionality.htm.) In your thesis statement (which must be *underlined*), you will develop an argument/opinion/primary insight *about* the specific topic/theme. In the rest of the essay, you must use at least ten assigned readings to illustrate and support your argument. You must turn in a rough draft of at least five full pages on Week 13. The rough draft is worth up to 75 points, based on both content (50%) and form (50%). When you turn in the final draft, you must also turn in the rough draft with my comments. The final essay is worth up to 150 points, based on both content (50%) and form (50%). (I reserve the right to award additional points for exceptional work.) Possible topics include: identity categories, identity politics, versions of feminism, transformation, coalition building, agency, survival, resistance, intervention, forging commonalities, "theory in the flesh," strategies for achieving solidarity, or transformation. Papers will be scored according to the following criteria:

Is the thesis statement sophisticated and well-developed but appropriate for the page length?
How closely does the essay adhere to the thesis statement?
How well does the essay follow the "Guidelines for Written Work"?
Do the selected texts appropriately and effectively support the thesis?
Does the paper successfully demonstrate the ways identities intersect with and impact thematic issues?

Late essays will be marked down, 10 points per day.

Extra Credit Opportunity. You may earn up to 25 points by reading and writing an analysis of Howard Zinn's *A People's History of the United States*, and you may earn up to 15 points by commenting on your peers' analyses of Zinn's book. Analyses must be posted by 10 p.m. on 12/2, and responses must be posted by 10 p.m. on 12/5. Your analysis should be a substantive engagement with Zinn's book, addressing the following questions: What were your reactions to Zinn's version of U.S. history? In what ways was it similar to the versions of history you learned in school (including college-level courses) and elsewhere (family, media, etc.)? In what ways was it different? In what ways (if any) did reading Zinn's book change your view of the United States and of history more generally? What did you find most compelling about *A People's History*, and what made it compelling? (Please do not just structure your analysis as a list of answers to these questions.) Your analysis should be at least 750 words in length. Each response should be at least 400 words and should engage with both your peer's analysis and Zinn's text.

Grade: A   B   C   D   F
Points: 598 or above   597–533   532–468   467–403   402 or below

Calculation of Points

Attendance	16×15	240 points
Discussion		50 points
Discussion questions	6×5	30 points
Talking Notes #1–4	4×15	60 points
Talking Notes #5		25 points
Intersectionality essay (rough draft)		75 points
Intersectionality essay (final draft)		150 points
Paper proposal		10 points
Thesis statement and introduction		10 points
Total		650 points

Course Expectations and Format[*]

Written assignments should be typed, double-spaced with 1-inch margins and 12-point font. (Talking Notes may be single-spaced with 11-inch Arial font.) Indent each paragraph, and do not add extra spaces between paragraphs. Number each page of your writing assignment. For more information see "Guidelines for Written Work" later in this syllabus.

Proofread Carefully. I expect you to revise and carefully proofread all assignments, including Blackboard postings. Your postings should be free of typos and grammatical errors.

Readings. All readings assigned for a specific date must be completed by class time on the date listed in the syllabus. I expect you to read the material thoughtfully and in an engaged manner (take notes, reflect on the material, etc.). Please read all endnotes and footnotes.

---

[*] Thanks to Dr. Claire Sahlin for allowing me to borrow from and modify portions of the following information from her WS 5663-01 Summer 2005 syllabus.

Talking Notes Selection

Talking Notes #1: This Bridge Called My Back

Mitsuye Yamada: "Asian Pacific American Women and Feminism" (74–79)

Judit Moschkovich: " 'But I Know You, American Woman' " (83–89)

Beverly and Barbara Smith: "Across the Kitchen Table" (123–40)

Andrea Canaan: "Brownness" (260–66)

Pat Parker: "Revolution" (267–72)

Merle Woo: "Letter to Ma" (155–63)

Gloria Anzaldúa: "OK Momma, Who the Hell Am I? An Interview with Luisah Teish" (247–59)

Norma Alarcon: "Chicana Feminist Literature: A Re-vision through Malintzin" (202–11)

Talking Notes #2: this bridge we call home, Sections 1 and 2

Renee M. Martinez: "Del puente al arco iris . . ." (42–50)

Alicia Rodriguez and Susana Vasquez: "Engaging contradictions, creating home" (53–59)

Helen Johnson: "Bridging Different Views: Australian and Asia-Pacific Engagements . . ."

Mita Banerjee: "The Hipness of Mediation" (117–25)

Shefal Milczarek-Desai: "Living Fearlessly with and within Differences" (126–35)

Marla Morris: "Young Man Popkin: A Queer Dystopia" (137–45)

Nathalie Handal: "Shades of a Bridge's Breath" (158–64)

Reem Abdelhadi and Rabab Abdulhadi: "Nomadic Existence" (165–75)

Susan Guerra: "IN THE END (AL FIN) WE ARE ALL CHICANAS" (181–90)

Talking Notes #3: this bridge we call home, Sections 3 and 4

Shirley Geok-lin Lim: "Memory and the New-Born" (208–23)

Nada Elia: "The 'White' Sheep of the Family: But Bleaching Is Like Starvation" (223–31)

Simona Hill: " 'All I Can Cook Is Crack On a Spoon': A Sign for a New Generation of Feminists" (258–66)

Donna Tanigawa: "Premature" (267–77)

Judith Witherow: "Yo' Done Bridge is Fallin' Down" (287–93)

Anonymous: "For My Sister: Smashing the Walls of Pretense and Shame" (295–301)

Renae Bredin: "So Far from the Bridge" (325–29)

Rosa Maria Pegueros: "The Ricky Ricardo Syndrome" (330–35)

Talking Notes #4: this bridge we call home, Sections 5 and 6

tatiana de la tierra: "Aliens and Others . . ." (358–68)

Kimberly Springer: "Being the Bridge" (381–89)

Jid Lee: "The Cry-Smile Mask" (397–402)

Cynthia Franklin: "Recollecting this Bridge in an Anti-Affirmative Action Era" (415–33)

Elana Dykewomon: "The Body Politic? Mediations on Identity" (450–58)

Diane Courvant: "Speaking of Privilege" (458–63)

Reyes: "The Latin American and Caribbean Feminist/Lesbian Encuentros" (463–70)

Ari Lev: "Tenuous Alliance" (473–82)

Talking Notes #5: Pedagogies of Crossing

1. "Erotic Autonomy as a Politics of Decolonization: Feminism, Tourism, and the State in the Bahamas" (21–65)
2. "Imperial Desire/Sexual Utopias: White Gay Capital and Transnational Tourism" (66–90)
3. "Whose New World Order?: Teaching for Justice" (91–116)
4. "Anatomy of a Mobilization" (117–80)
5. "Transnationalism, Sexuality, and the State: Modernity's Traditions at the Height of Empire" (181–254)
6. "Pedagogies of the Sacred: Making the Invisible Tangible" (287–32)

*Schedule of Class Meetings**

Week 1 Introduction: Course expectations; a brief history of U.S. racialization and Native peoples.

Week 2 *Read*: Winona LaDuke: *Last Standing Woman* (1–155); Howard Zinn: *People's History* (1–22; 125–48).

---

* The following is a *tentative* schedule; I might change readings, due dates, or assignments, so please be sure to check Blackboard for any schedule changes.

Week 3 *Read*: LaDuke: *Last Standing Woman* (156–end); Zinn: *People's History* (503–40; 601–30).

Note: Before today's class please look through both bridge books and select several essays that you would like to do for Talking Notes #1–4.

Week 4 *Read*: Octavia Butler: *Kindred*; Zinn: *People's History* (23–38; 39–58).

Week 5 *Read*: Beth Roy: "For White People, on How to Listen When Race is the Subject" (HO); Kimberlé Crenshaw: "Demarginalizing the Intersection of Race and Sex" (HO); *This Bridge Called My Back*: Toni Cade Bambara: "Foreword, 1981" (xl–xliii); Moraga: "Preface, 1981" (xliv–li); Moraga and Anzaldúa: "Introduction, 1981" (lii–lvi); Kate Rushin: "The Bridge Poem" (lvii–lviii); Moraga: "Refugees Of A World On Fire, Foreword to the 2nd edition, 1983" (346–50); Anzaldúa: "Foreword to the Second Edition" (351–52); Moraga: *"Children Passing in the Streets: The Roots of Our Radicalism"* (3–4); Nellie Wong: "When I Was Growing Up" (5–6); mary hope lee: "on not bein' " (7–9); Moraga: "For the Color of My Mother" (10–11); Rosario Morales: "I Am What I Am" (12–13); Chrystos: "He Saw" (16–17); "Entering the Lives of Others: Theory in the Flesh" (21–60).

Week 6 *Due*: Talking Notes #1.

Read: *This Bridge Called My Back*: Moraga: "And When You Leave . . . : Racism in the Women's Movement" (63–65); Jo Carillo: "And When You Leave, Take Your Pictures with You" (66–67); doris davenport: "The Pathology of Racism" (90–96); Morales: "We're All in the Same Boat" (97–100); Audre Lorde: "An Open Letter to Mary Daly" (101–05); Lorde: "The Master's Tools Will Never Dismantle the Master's House" (106–09); Moraga: "Between the Lines: On Culture, Class, and Homophobia" (113–15); Cheryl Clarke: "Lesbianism: An Act of Resistance" (141–51); Mirtha Quintanales: "I Paid Very Hard for my Immigrant Ignorance" (167–74); Anzaldúa: "Speaking in Tongues: The Third World Woman Writer" (181–82); Anzaldúa: "Speaking in Tongues: A Letter to Third World Women Writers" (183–93); Nellie Wong: "In Search of the Self as Hero" (196–201).

Week 7   *Read*: *This Bridge Called My Back*: Anzaldúa: "El Mundo Zurdo: The Vision" (215–16); Anzaldúa: "La Prieta" (220–33); Combahee River Collective: "A Black Feminist Statement" (234–44); Moraga: "The Welder" (245–46); Moraga: "From Inside the First World Foreword, 2001" (xv–xxxiii); Anzaldúa: "Foreword, 2001" (*Bridge* xxxiv–xxxix). In *this bridge we call home*: Anzaldúa: "(Un)natural bridges, (Un)safe spaces" (1–5); AnaLouise Keating: "Charting Pathways, Marking Thresholds . . . A Warning, An Introduction" (6–20); Chela Sandoval: "AFTERBRIDGE: Technologies of Crossing" (20–26).

Week 8   *Due*: Talking Notes #2.

   Read: Iobel Andemicael: "Chameleon" (28–42); Hector Carbajal: "Nacido En Un Puente/Born on a Bridge" (50–52); Jesse Swan: Bridges/Backs/Books: A Love Letter to the Editors" (59–62); Rebecca Aanerud "Thinking again: This Bridge . . . and the Challenge to Whiteness" (69–77); Donna Langston: "The Spirit of *This Bridge*" (77–81); Jacqui Alexander: "Remembering *This Bridge*, Remembering Ourselves" (81–103); Joanne DiNova: "Seventh Fire" (104); Evelyn Alsutany: "Los Intersticios: Recasting Moving Selves" (106–10); Leticia Hernández-Linares: "Gallina Ciega: Turning the Game on Itself" (110–16); Berta Avila: "THE REAL" (116–17); Carbajal: "A Letter to a Mother, from Her Son" (136–37); Jody Norton: "Transchildren, Changelings, and Fairies" (145–54); Kim Roppolo: "The Real Americana" (155–58); Minh-ha Pham: "(Re)Writing Home: A Daughter's Letter to Her Mother" (176–80).

Week 9   *Due*: Talking Notes #3.

   Read: Deborah Miranda: "What's Wrong with a Little Fantasy?" (192–20); Miranda and Keating: "Footnoting Heresy: Email Dialogues" (202–08); Clarke: "Lesbianism, 2000" (232–39); Valerio: "Now That You're a White Man" (239–54); mary loving blanchard: "Poets, Lovers, and the Master's Tools" (254–57); Marisela B. Gomez: "Council Meeting" (293–94); Nadine Naber: "Resisting the Shore" (301–03); Chandra Ford: "Standing on This Bridge" (304–13); Genny Lim: "Stolen Beauty" (313); Joann Barker: "Looking for Warrior Woman (Beyond Pocahontas)" (314–25); Jeanette Aguilar: "Survival" (339–41); Cervenak et al.: "Imagining Differently: The Politics of Listening in a Feminist Classroom" (341–56).

Week 10 *Due*: Talking Notes #4.

Read: Sunu P. Chandy: "The Fire in My Heart" (369–71); Laura Harris: "Notes from a Welfare Queen in the Ivory Tower" (372–81); Bernadette García: "This World Is My Place" (390); Mirtha Quintanales: "Missing Ellen and Finding the Inner Life" (391–96); Toni King et al. "Andrea's Third Shift" (403–15); Irene Lara: "Healing Sueños for Academia" (433–38); Carmen Morones: "The Colors Beneath Our Skin" (440–48); Maria Proitsaki: "Connection: The Bridge Finds Its Voice" (449–50); Ednie Garrison: "Sitting in the Waiting Room of Adult and Family Services . . ." (470–72); Alicia Gaspar de Alba: "Chamizal" (483–85); Indigo Violet: "Linkages" (486–94).

Week 11 *Read*: *this bridge we call home*: section vii. " 'i am the pivot for transformation' . . . enacting the vision" (495–578).

Due: Paper proposal: Your proposal (typed, double-spaced) should include the following: (1) In one paragraph of at least five sentences, discuss your topic and your tentative argument. (2) A list of at least five pieces you might use. Please bring four copies. (Worth up to 10 points, based on both content and form.)

Week 12 *Read*: Alexander: *Pedagogies of Crossing*: "Introduction" (1–20).

Due: Tentative Thesis Statement and introductory paragraph for Intersectionality Essay (must be typed). Please bring two extra copies. (Worth up to 10 points.)

Note: Before today's class please look through *Pedagogies of Crossing* and select two chapters that you'd like to use for Talking Notes # 5. Your selection should not include chapter 6.

Week 13 *Due*: Rough draft (must be at least five full pages, typed, double-spaced. Must use at least seven course texts.) Please bring three extra copies.

Week 14   *Read*: Mariana Ortega: "Being Lovingly, Knowingly Ignorant: White Feminism and Women of Color" (HO).

Due: Talking Notes #5.

Week 15   *Due*: Blackboard Activity (TBA).

Week 16   *Due*: Intersectionality Essay (*note*: please turn in both the final copy and the rough draft w/ my comments.)

Platica.

★   ★   ★

Transgressive Identities: Queer Theories/Critical 'Race' Theories

WS 5663 50% on-line via Blackboard
Dr. AnaLouise Keating

> **trans.gress** Function: *verb* Etymology: Middle French *transgresser*, from Latin *transgressus*, past participle of *transgredi* to step beyond or across, from *trans-* + *gradi* to step Date: 1526
>> *transitive senses* **1** : to go beyond limits set or prescribed by : **VIOLATE** <*transgress* divine law> **2** : to pass beyond or go over (a limit or boundary) *intransitive senses* **1** : to violate a command or law : **SIN 2** : to go beyond a boundary or limit
>> **trans.gres.sive** / *adjective*
>> **trans.gres.sor** / *noun*
>>   Merriam-Webster's Online Dictionary, 10th ed.

Language and testimony are both means of critically and vigilantly interpreting and reinterpreting our relationships and the boundaries of ourselves, and the means of circumscribing our limits and performing the labor of the negative. The relation between love and language brings with it a strong ethical obligation to speak and to listen to others. The obligation to speak and to listen opens a public space for love. And, it is this public space that makes love ethical and political.

                                                            Kelly Oliver

Sticks and stones may break our bones, but words—words that evoke structures of oppression, exploitation, and brute physical threat—can break souls.

                                                  Kwame Anthony Appiah

For me there aren't little cubbyholes with all the different identities—intellectual, racial, sexual. It's more like a very fine membrane—sort of like a river, an identity is sort of like a river. It's one and it's flowing and it's a process. By giving different names to different parts of a single mountain range or different parts of a river, we're doing that entity a disservice. We're fragmenting it. I'm struggling with how to name without cutting it up.

Gloria E. Anzaldúa

Do I contradict myself? / Very well then I contradict myself, / (I am large, I contain multitudes.)

Walt Whitman

Course Description

This reading-intensive, interdisciplinary seminar explores feminist/womanist interventions into recent developments in Critical 'Race' Theories and Queer Theories, focusing primarily on the issue of passing (racial/sexual/gender impersonation, masquerade, and drag). Through readings, film viewings, and discussion, we will examine some of the ways passing challenges, deconstructs, and/or reinforces stable identity categories and social formations.

Course Goals/Student Learning Outcomes

Students who successfully complete this course will obtain the following: (1) An increased understanding of feminist/womanist theorizing, critical 'race' studies, queer theories, models of identity formation, the history of racialized identity formation in the United States, the relationship between discursive and material interventions into existing social and identity structures; (2) an increased ability to think relationally and critically; (3) an increasingly nuanced understanding of the ways commonalities, similarities, and differences work together.

Required Texts

- Richard Delgado and Jean Stefancic: *Critical Race Theory: An Introduction*
- Donald E. Hall: *Queer Theories* (cited as QT in syllabus)
- Langston Hughes: *The Ways of White Folks* (cited as WWF in syllabus)
- Nella Larson: *Quicksand and Passing*

- Joan Nestle, Riki Wilchins, Clare Howell: *GenderQueer: Voices from Beyond the Sexual Binary* (cited as GQ in syllabus)
- Carol Queen and Larry Schimel: *Pomosexuals: Challenging Assumptions about Gender and Sexuality*
- George Schuyler: *Black No More*
- Thandeka: *Learning to Be White: Money, Race, and God in America*
- Patricia Williams: *Alchemy of Race and Rights: Diary of a Law Professor*
- Valdes, McCristal, and Harris: *Crossroads, Directions, and a New Critical Race Theory* (cited as CD and NCRT)

All books are on reserve at the Blagg-Huey Library.

Recommended: Beidler and Taylor: *Writng Race across the Atlantic World.*

Films (might include): *Illusions*; *The Question of Equality*; *Looking for Langston*; *Paris Is Burning*; *You Don't Know Dick.*

Course Packet (CP)

Mary Coombs: "LatCrit Theory and the Post-Identity Era: Transcending the Legacies of Color and Coalescing a Politics of Consciousness;" Adrian Piper: "Passing for White, Passing For Black;" David Roediger: "The White Question;" Lisa Duggan: "Making It Perfectly Queer;" Henry Louis Gates, Jr.: "Looking for Modernism;" Richard Dyer: "White;" John Hartigan Jr.: "Establishing the Fact of Whiteness;" George Lipsitz: "The White 2K Problem;" Walter O. Bockting and Charles Cesaretti: "Spirituality, Transgender Identity, and Coming Out;" Richard A. Posner: "Narrative and Narratology in Classroom and Courtroom;" Derrick Bell: "The Power of Narrative;" Daniel G. Solórzano and Tara J. Yosso: "Critical Race Methodology: Counter-Storytelling as an Analytical Framework for Education Research;" Richard Delgado: "Two Ways to Think About Race;" Angela P. Harris: "Building Theory, Building Community;" Katrina Roen: "Transgender Theory and Embodiment;" Richard Delgado: "Derrick Bell's Toolkit: Fit to Dismantle That Famous House?"; George Schuyler: "Our Greatest Gift to America;" Judith Butler: "Gender Is Burning;" bell hooks: "Is Paris Burning?;" Elisa Glick: "Sex Positive: Feminism, Queer Theory, and the Politics of Transgression;" Paul Gilroy: "Race Ends Here;" James Baldwin: "On Being 'White' and Other Lies;" Fran Peavey: "Strategic Questions Are Tools for Rebellion;" Charlotte Perkins Gilman: "The Yellow Wall-Paper."

BB Course Documents (CD)

john a. powell: "Whiteness: Some Critical Perspectives: Dreaming of a Self Beyond Whiteness and Isolation;" john a. powell: "Whites Will Be Whites: The Failure to Interrogate Racial Privilege."

As You Read. WS 5663 is a graduate-level course, and I expect all students to follow graduate-level academic practices: (1) I expect you to read the material thoughtfully and in an engaged manner (underline/take notes, reflect, reread, etc.); (2) I expect you to read all endnotes and footnotes; and (3) I expect you to seek out definitions for terminology you don't know. This reading-intensive course takes a dialogic approach. We'll be asking the following questions, so please think about them as you read: How would you describe the relationship between language, personal/collective identities, and material reality? What impact can storytelling and other forms of language have on material reality (that is, on the ways we see ourselves and others, live our lives, and act)? What roles can storytelling play in changing individual and collective consciousness and in other ways bringing about progressive social change? What is the relationship between self and other? How can CRT (Critical Race Theory) and QT (Queer Theory) be used in the service of social justice?

Grades

Grades will be awarded on a point system. *It is your responsibility to keep track of your number of points.* Please do not e-mail or call me at the end of the semester to find out your grade.

Attendance. Our class meets in person five times during the semester on Monday nights. Each class is worth 30 points. 10 points will be sub-tracted for late arrivals and/or early departures. If you miss more than one class period, the highest grade you can receive in this course is a B (unless, in very exceptional cases, you have spoken with me in advance and complete additional work). If you must miss class, please notify me ahead of time, if possible, by e-mail. It is your responsibility to find out (from your classmates!) about announcements, syllabi change, handouts, etc.

Participation. Participation entails engaged, respectful listening, speaking, and responding to postings. In an on-line environment, "listening" also include reflective reading. Please do not be too rushed in your

Blackboard replies; take time to think about your fellow students' postings. Your responses should be thoughtful, thought-provoking, and well-written. I have set up, on the Discussion Board, a Forum called *EXPRESS YOURSELF!* I hope that you will use this forum to share your ideas, thoughts, reflections, and so on about the issues raised in this course. Occasionally I might suggest topics for discussion. (In-class participation and replying to Discussion Board responses to your postings is worth up to 50 points.)

In-Class and Blackboard Discussion Questions. Each week, you will design three discussion questions (a question you have about one or more of the assigned readings that you would like us to explore in class). *One* question will be due at the beginning of most in-class sessions; this question must be typed, double-spaced, 12-inch font. Generally, *two* questions will be due on the Discussion Board Wednesdays by 9 p.m. A discussion question can deal specifically with one of the assigned readings or it can be a bit broader and engage several readings. (The two discussion questions posted on BB should not both focus on the same text.) Discussion questions will significantly guide our class conversations and serve at least three additional purposes:

- They offer opportunities for students to reflect more deeply on the assigned readings and, through this reflection, to deepen their learning.
- They enhance student accountability and give students more control over our time together.
- They allow me to assess student interests and learning.

Question Format Instructions. If you ask a question about a specific passage, please provide the quotation and page number(s); if you ask a question about an issue found on specific pages, please include the page numbers. (There is no need to provide full bibliographical information for these questions.) Mark with an asterisk (★) those questions that you really hope we'll have time to discuss. Do not ask definition-based or other easily researched questions (for instance: "What does 'transgender' mean?" or "What is a mulatto?") Each discussion question is worth up to 10 points (evaluated on both content and form). *No points will be given for late discussion questions.*

Instructions for Posting BB Discussion Questions. Each week I will set up a forum on the Discussion Board for that week's questions. Begin a separate thread for each of your discussion questions. In the subject heading, include a title that informs us about your question. (For instance: "Passing in Hughes's "Rejuvenation by Joy," or "Identity in Schuyler and Thandeka.")

Discussion Question Responses. Each week you are required to respond to a specific number of your peers' discussion questions for that week. All responses are due by Friday at 9 p.m. of that week. Each response is worth up to 5 points, based on both content and form. Although you will not receive extra points for additional responses, additional responses and replies contribute to your participation grade. *No credit will be given for late Discussion Board responses!*

Talking Notes Handouts. You will select and read an essay from the following assigned texts, prepare a short handout, and make a short (five minutes) presentation:

> *GenderQueer; Pomosexuals; Crossroads, Directions, and a New Critical Race Theory* (extra credit course packet TN described below.)

For the in-class Talking Notes (6/26 and 7/3), you will need to bring copies of your handout for the entire class. (You may use the copier in the Women's Studies office.) For further instructions see pages 6 and 7 of this syllabus. To receive full credit for TN on *Pomosexuals* and *Crossroads*, you must be in class on the evening they are due. These Talking Notes handouts serve several purposes, including the following: (1) they allow us, as a class, to cover more material; (2) they enable students to work on critical thinking, reading, and writing skills. Talking Notes for *Pomosexuals* and *GenderQueer* are worth up to 10 points; Talking Notes for *Crossroads* and the course packet are worth up to 25 points. All points are based on content, form, presentation, and text selection. You will choose texts in class on 6/12.

Final Project. A self-reflective essay (five to ten pages, double-spaced) exploring what you've learned in "Transgressive Identities." (Of course, I hope you've learned so much that *ten* pages seems too short. If so, be selective: discuss the most important insights you have acquired.) Some areas you might discuss include: identity formation, identity labels, QT, CRT, theory and theorizing, LGBT identities, "race," "whiteness," critical 'race' studies and queer studies as interdisciplinary fields; the history of racialized identity formation in the United States; the relationship between discursive and material interventions into existing social and identity structures; the ways commonalities, similarities, and differences work together. During the course, we'll have in-class writings that should be useful as you develop this essay.

Suggestions. Reflect on the course content (discussions, readings, in-class writings, your journal entries, and films) and chart the shifts in your views. After reflecting on the course, develop a thesis statement, which can be something simple—along the lines of: "Before taking this course, I believed ——; however, I now recognize ——;" or "In 'Transgressive Identities' I learned X things;" or "I learned a number of things in this course, most importantly ——;" If, in fact, you learned nothing in this course, you could have a thesis statement like the following: "While I learned nothing new in 'Transgressive Identities,' I was especially struck by ——;" I would really appreciate it if you'd answer honestly, rather than attempt to write something that you believe I'd like to hear. In other words: Don't write to please me! Write to reflect on what you've really learned. I realize that this type of assignment is tricky, because of the immense temptation to resort to flattery. However, it's worth the risk because self-reflective essays can also be a useful self-learning tool for students.

Worth up to 100 points, based on both content and form. I reserve the right to award extra points for exceptional work. (By "content," I mean how well you follow and support your thesis statement—not how much I "like" what you say!) Please follow the "Guidelines for Written Work."

Extra Credit. You may earn up to 25 points by completing a Talking Notes on any of the course packet recommended readings (see below for a list), and you may earn up to 15 points by commenting on any Talking Notes posted to this site. Your Talking Notes must be posted to the "Extra Credit" Discussion Board Forum by 9 p.m. on the date each recommended reading is listed in the syllabus. (For directions on Talking Notes format see below.) Each response to these posted Talking Notes will be worth up to 5 points based on both content and form.

Grade:	A	B	C	D	F
Points:	540–496	495–442	441–388	387–334	333 or below

Calculation of Points

Attendance	5×30	150
Participation		50
Discussion questions (in-class)	3×10	30
Discussion questions (BB)	8×10	80
Discussion Board responses	17×5	85
Talking Notes: *GenderQueer*		10
Talking Notes: *Pomosexuals*		10
Talking Notes: *Crossroads*		25
Final		100
Total points possible		540

*Course Expectations and Format**

Deadlines. In the interest of fairness toward students who meet deadlines *and* to ensure that you can complete all coursework within the five-week semester, all due dates (for discussion questions, Discussion Board postings, final essay, etc.) are firm and will not be extended unless an actual emergency prevents a student from conforming to them. Technical difficulties, requirements of other courses and of your employment, and other events that could have been foreseen, do not count as emergencies under this policy. Requests for extensions must be made before the announced dates unless it is actually impossible for you to do so.

Written assignments. These should be typed, double-spaced with 1-inch margins and 11- or 12-point font. Indent each paragraph, and do not add extra spaces between paragraphs. Number each page of your writing assignment. For more information see "Guidelines for Written Work" later in this syllabus.

Proofread Carefully. I expect you to revise and carefully proofread all Blackboard postings. Your postings should be free of typos and grammatical errors.

Readings. All readings assigned for a specific date must be completed by class time on the date listed in the syllabus. I expect you to read the material thoughtfully and in an engaged manner (take notes, reflect on the material, etc.). Please read all endnotes and footnotes.

Due Dates. Papers and Blackboard assignments are due on the date and time listed.

Guidelines for "Talking Notes" Handouts/Discussions

As explained above, for several classes you will be required to prepare a summary and handout of selected articles. Please do not exceed the maximum page length (see below). *Do not try to write a short essay.* Instead, use bullets and, where appropriate (for example, in the summary), paragraphs. Please be sure to include your text's title, author, and full page numbers for the essay. For example: Adrian Piper: "Passing for White, Passing For Black" (CP 5–20).

Talking Notes for Narratives in GenderQueer and Pomosexuals. Your handout should be only one page. (To maximize the page, you may single space

---

* Thanks to Dr. Claire Sahlin for allowing me to borrow from and modify portions of the following information from her WS 5663-01 syllabus.

and use 11-inch Arial font). Please include the following information and bolded headings in your Talking Notes:

- *Summary*: Summarize the narrative, including the main points you believe the author to be making.
- *Transgressions*: Discuss the ways that this author seems to transgress conventional identity categories.
- *Exemplary Quotations*: Select one or two key quotations that seem to best illustrate the article. (During your discussion, be prepared to explain why they illustrate the article.)
- *Related Course Readings*: What other readings from this course does this piece resemble? (During your discussion, explain why.)

Talking Notes for the Essays in Crossroads *and Course Packet.* (The latter is optional extra-credit.) Your handout may be no longer than two pages. (To maximize the page, you may single space and use 11-inch Arial font). Please include the following information and bolded headings in your Talking Notes:

- *Summary*: Summarize the essay, including the main argument and the writing style (storytelling, conventional academic style, hybrid, etc.). (In your discussion, be sure to explain why you did or did not find the argument persuasive.)
- *Outline*: Outline the essay structure (include page numbers).
- *Exemplary Quotation(s)*: Select one or two key quotations that seem to best illustrate the article. (During your discussion, be prepared to explain why you believe that they illustrate the article.)
- *Related Course Readings*: What other readings from this course does this piece resemble? (During your discussion, explain why.)

Crossroads text selection: Select a chapter from Part II, "Crossroads."

*Schedule for Class Meetings and Activities**

Week 1   M *Meet* in HDB 300 from 6 to 9:50 p.m.

Introduction: Student introductions, course overview, terms and definitions.

---

* The following is a *tentative* schedule. I might change readings, due dates, or assignments. Please check BB and your e-mail daily, and e-mail me if you miss class. Always check BB for changes in specific assignments.

Film: *A Question of Equality*; *Illusion*.

W *Due* by 9 p.m. tonight: Two discussion questions over the following readings: Roediger: "The White Question" (CP 21–23); Nestle, Wilchins, Howell: "Introductions: Three Voices" (GQ 3–22); Wilchins: "It's Your Gender, Stupid!" (GQ 23–32); Queen and Schmiel: "Introduction" (*Pomosexuals* 19–25); Hall: "Introduction" (QT 1–19); Gilroy: "Race Ends Here" (CP 237–46); Baldwin: "On Being 'White' and Other Lies": (CP 247–49); Fran Peavey: "Strategic Questions Are Tools for Rebellion" (CP 250–61); Matsuda: "Beyond, and Not Beyond, Black and White: Deconstruction Has a Politics" (CD and NCET 393–98).

F *Due* by 9 p.m. tonight: By this point, you should have responded to at least four of your classmates' discussion questions posted for this week. (Each response should be at least 100 words.)

Week 2 M *Meet* tonight in HDB 300 from 6 to 9:50 p.m.

Due at 6 p.m. tonight: One discussion question over one or more of the following readings: Delgado and Stefancic: *Critical Race Theory*; Bell: "The Power of Narrative" (CP 86–112); Gilman: "The Yellow Wall-Paper" (CP 262–274); Chang: "Critiquing 'Race' and Its Uses" (CD and NCRT 87–96); Montoya: "Celebrating Racialized Legal Narratives" (CD and NCRT 243–50); Romany: "Critical Race Theory in Global Context" (CD and NCRT 303–09); Jacob Hale: "Suggested Rules for Non-Transsexuals Writing about Transsexuals, Transsexuality, Transsexualism, or Trans ——": http://sandystone.com/hale.rules.html.

Film: *You Don't Know Dick*.

Also Due: Come prepared to make your selections for Talking Notes on *GenderQueer*, *Pomosexuals*, and *Crossroads, Directions, and a New Critical Race Theory*. (I recommend that you come with three possible choices for each text.)

W *Due* by 9 p.m. tonight: Talking Notes on *GenderQueer*. Two discussion questions over the following readings: Hall: "The Social Construction of Queer Theories"

(QT 12–111); Hall: "The Queerness of 'The Yellow Wallpaper' " (QT 115–29).

Recommended Readings: Bockting and Cesaretti: "Spirituality, Transgender Identity, and Coming Out" (69–78); Glick: "Sex Positive: Feminism, Queer Theory, and the Politics of Transgression" (CP 209–36).

F   *Due* by 9 p.m. tonight: Respond to at least two of your classmates' discussion questions and two Talking Notes posted for this week. (Each response should be at least 100 words.)

Week 3   M   *Meet* tonight from 6 to 9:50 p.m.

Due at 6 p.m. tonight: One discussion question over one or more of the following readings: Thandeka: *Learning to Be White*; Hughes: "Cora Unashamed" (WWF 3–18); Hughes: "Rejuvenation through Joy" (WWF 69–98); Lipsitz: "The Y2K White Problem" (CP 61–68).

Recommended Readings: Hartigan: "Establishing the Fact of Whiteness" (CP 51–61).

Film: *Paris Is Burning*.

W   *Due* by 9 p.m. tonight: Two discussion questions over the following readings: Schuyler: *Black No More*; "Our Greatest Gift to America" (CP 185–90).

F   *Due* by 9 p.m. tonight: Respond to at least three of your classmates' discussion questions posted for this week. (Each response should be at least 100 words.)

Week 4   M   *Meet* tonight from 6 to 9:50 p.m.

Due at 6 p.m. tonight: One discussion question over one or more of the following readings: Coombs: "LatCrit Theory and the Post-Identity Era" (CP 1–4); Duggan: "Making It Perfectly Queer" (CP 24–34); Christina: "Loaded Words" (*Pomosexuals* 29–35); Dyer: "White" (CP 39–50); Hughes: "Passing" (WWF 51–57); Bornstein: "Queer Theory and Shopping" (*Pomosexuals* 13–17); john a. powell: "Whiteness: Some Critical Perspectives: Dreaming of a Self Beyond Whiteness and Isolation" (CD).

Due: Talking Notes handout on *Pomosexuals*.

Recommended Readings: Solórzano and Yosso: "Critical Race Methodology: Counter-Storytelling as an Analytical Framework for Education Research" (113–24).

W   *Due* by 9 p.m. tonight (Wed.): Two discussion questions over the following readings: Riki Wilchins: Three essays (GQ 33–63); hooks: "Is Paris Burning?" (CP 202–08); Butler: "Gender Is Burning" (CP 191–201); Patricia Williams: *Alchemy of Race and Rights* (1–51) and one additional chapter of your choice; Posner: "Narrative and Narratology in Classroom and Courtroom" (CP 79–85).

Recommended Readings: Delgado: "Two Ways to Think about Race" (CP 125–43); Delgado: "Derrick Bell's Toolkit: Fit to Dismantle that Famous House?" (CP 167–84).

F   *Due* by 9 p.m. tonight: Respond to at least three of your classmates' discussion questions posted for this week. (Each response should be at least 100 words.)

Week 5 M   *Meet* tonight (Monday) in HDB 300 from 6 to 9:50 p.m.

Read: Larsen: *Passing*; Gates, "Looking for Modernism" (CP 35–38). (*Note*: No discussion question tonight!)

Due: Talking Notes on *Crossroads, Directions, and a New Critical Race Theory*.

Recommended Readings: Piper: "Passing for White, Passing for Black" (CP 5–20); Harris: "Building Theory, Building Community" (CP 144–55); Roen: "Transgender Theory and Embodiment: The Risk of Racial Marginalisation" (CP 156–66); john a. powell: "Whites Will Be Whites: The Failure to Interrogate Racial Privilege" (CD).

Film: *Looking for Langston*.

W   *Due*: by 9 p.m. tonight you should have done the following: (1) e-mailed your final essay to me; (2) posted a summary of your final essay (300–500 words) on Discussion Board.

F   *Due*: by 9 p.m. tonight: three responses (at least 100 words) to two of your classmates' final essay summaries.

NOTES

Introduction: Transformational Multiculturalism

1. M. Jacqui Alexander, "Remembering *This Bridge*, Remembering Ourselves: Yearning, Memory, and Desire," 85.
2. "Nepantlera" is a term Anzaldúa coined to describe "spiritual activists" who work for social justice by mediating among diverse groups. See her essay "now let us shift" for an extensive discussion of nepantleras. Also note that, like Anzaldúa, I choose not to italicize non-English words because to do so denormalizes and otherizes them.
3. In her discussion of scholarship linking spirituality with sociopolitical change Alexander refers to work by Gloria Anzaldúa, Cornel West, bell hooks, Lata Mani, and Leela Fernandes.
4. I am talking here about spiritual activism—a visionary, experience-based epistemology, politics, and ethics, a way of life and a call to action. For an extensive discussion of spiritual activism see my essay "Shifting Perspectives: Spiritual Activism, Social Transformation, and the Politics of Spirit." For additional examples of how scholars incorporate politically engaged forms of spirituality into their scholarship, see john a. powell's "Does Living a Spiritually Engaged Life Mandate Us to Be Actively Engaged in Issues of Social Justice?" and Cynthia Dillard's *On Spiritual Strivings*.
5. See Owen J. Dwyer and John Paul Jones III's discussion of "categorical naturalization" (212).
6. Throughout *Teaching Transformation* I use scare quotes to mark the word *race*. I do so to de-normalize and de-naturalize the term. I understand that the repeated use of scare quotes might be frustrating for some readers, but this frustration is part of my point! For more on this topic, see my comments in this premise and in chapter three.
7. See Lorde's essay "The Master's Tools Will Not Dismantle the Master's House" in her *Sister Outsider*. To be fair, I should acknowledge that Lorde was not referring specifically to 'race' when she referred to "the master's tools."
8. See Kwame Anthony Appiah's "The Conservation of 'Race'" and "The Uncompleted Argument: Du Bois and the Illusion of Race" as well as Naomi Zack's *Race and Mixed Race*.
9. See George Lipsitz's *The Possessive Investment in Whiteness*.
10. I mark the words *white* and *whiteness* with scare quotes to de-normalize and de-naturalize them. I discuss this issue in much greater depth in chapters three and four.
11. For more on the racism embedded within contemporary discussions of a "colorblind" U.S. society see the opening chapter in Patricia Hill Collins's *From Black Power to Hip-Hop: Racism, Nationalism, and Feminism*. There are a very few exceptions to conservative ('white') color-blind narratives. See Reginald Leamon Robinson's publications for an attempt to develop progressive color-blind approaches to social justice. See also Richard T. Ford's "Race as Culture? Why Not?" and cristie e. cunninham's "The 'Racing' Cause of Action and the Identity Formally Known as Race: The Road to Tamazunchale."

12. Although Mitchell and Feagin comment only on racial/cultural groups, I would suggest that their assertion could also apply to other oppressed groups, including (but not limited to) those marginalized by gender, sexuality, ability, health, religion, and so on.

13. Christa Downer discusses this limited effectiveness in her dissertation, *The Making of a Pluralistic Egalitarian Society: Reconceptualizing the Rhetoric of Multiculturalism for the 21st Century.*

14. I do not capitalize the words *western* and *eurocentric* because I feel that to capitalize them reinforces their existing dominant status.

15. For additional discussion of the ways oppositional consciousness and politics have been informed by eurocentric epistemologies, see Mark Lawrence McPhail's discussions of complicity in "Complicity: The Theory of Negative Difference" and "The Politics of Complicity: Second Thoughts about the Social Construction of Racial Equality." As he explains in the former, "To the extent that critics of race, gender, and language oppose hegemonic discourses based upon positions that subscribe to this rhetoric of negative difference, they become complicitous with those discourses, and in fact reify them" (4).

16. As Marshall B. Rosenberg, founder and educational director of the Center for Nonviolent Communication, notes, "Judgmental, or right/wrong thinking, is one of the hardest things I've found to overcome in teaching Nonviolent Communication over the years. The people that I work with have all gone to schools and churches and it's very easy for them, if they like Nonviolent Communication, to say it's the 'right way' to communicate."

17. See also john a. powell's discussion of the limits of oppositional politics in "Whites Will Be Whites" where he maintains that "[t]o the extent that [marginalized groups] remain only oppositional, or try to assimilate as an end, they also remain dependent on the dominant discourse that they seek to challenge" (434).

18. I use the phrase "conventional theory" to acknowledge the many forms theorizing can take. Like Barbara Christian, I believe that theory includes poetry, fiction, autobiography, and much more.

19. Thanks to Maria Elena Cepedes for introducing me to Frances Aparicio's work.

20. Laurie Grobman makes a similar point: "Over the last two decades, coincident with the broadening of the literary canon, multicultural scholars have produced a vast amount of critical and pedagogical literature. Despite these advances, though, and despite a broad consensus about the moral and political goals of our work in and out of the classroom . . . we lack a coherent pedagogy" (225). Maitino and Peck's anthology provides an important exception to this trend; each essay includes a section entitled "Teaching the Work."

21. As Henry Giroux states, "Multiculturalism covers an extraordinary range of views about the central issues of culture, diversity, identity, nationalism, and politics" ("Democracy and the Discourse of Cultural Difference" 507, n. 1).

22. See, for instance, Timothy Powell's assertion that "Multiculturalism, too often, has been co-opted by state and corporate narratives as a way to mask social inequalities with a rhetoric that celebrates cultural differences" (*Ruthless Democracy* 8).

23. As Eric Sundquist asserts, "[n]o matter the breadth or diversity of new formulations of the canonical tradition: . . . it remains difficult for many readers to overcome their fundamental conception of 'American' literature as solely Anglo-European in inspiration and authorship, to which may then be added an appropriate number of valuable 'ethnic' or 'minority' texts, those that closely correspond to familiar critical and semantic paradigms" (7). Gregory Jay makes an analogous point:

> Current revisionary critiques show that the "American" of conventional histories of American literature has usually been white, male, middle- or upper-class, heterosexual, and a spokesman for a definable set of political and social interests. Insofar as women, African- and Asian- and Native-Americans, Hispanics, gays and lesbians, and others make an appearance in such histories, it is usually in terms of their also being made into spokesmen for traditional values and schemes. Their assimilation into American literature comes at the cost of their

cultural heritage and obscures their real antagonism and historical difference in relation to the privileged classes. (267)

24. Jane Tompkins provides an excellent analysis of the nineteenth-century politics of publishing and canon formation in *Sensational Designs*.

25. See Trinh Minh-ha's critique of this type of multiculturalism, which she describes as "the juxtaposition of several cultures whose frontiers remain intact" (232).

26. This emphasis on narrowly defined difference also limits how canon formation occurs within ethnic-specific literatures. For more on this topic, see Gene Andrew Jarrett's discussions of how African-American literature has been defined too narrowly. In brief, he demonstrates that authors and books which do not focus on "black" life have been ignored. As he explains in "Judging a Book by Its Writer's Color,"

> This definition has long imposed a mythical "one-drop rule" on authors, meaning that one drop of African ancestral blood coursing through their bodies makes them black. It has also dictated African-American "canon formation," misleading readers into believing that black people write best only about black people.
>
> This authentic idea of African-American literature—perpetuated by publishers, acquisition editors, anthology editors, scholars, teachers, and ultimately students—is everywhere. It determines the way authors think about and write African-American literature; the way publishers classify and distribute it; the way bookstores receive and sell it; the way libraries catalog and shelve it; the way readers locate and retrieve it; the way teachers, scholars, and anthologists use it; and the way students learn from it. In short, it determines our belief that we supposedly know African-American literature when we see it.

For a concise summary of Jarrett's argument see his "Not Necessarily Race Matter."

27. See, for example, Cornel West on "[s]truggles over turf, slots, curriculum, and debates over multiculturalism" ("Eurocentrism and Multiculturalism" 120). See also Riki Wilchins' discussion of gender turf wars in *Read My Lips: Sexual Subversion and the End of Gender* (79–88).

28. On "revolutionary multiculturalism," see Peter McLaren's *Revolutionary Multiculturalism: Pedagogies of Dissent for the New Millennium*. For a useful summary of critical multiculturalism, see Henry A. Giroux: "Critical multiculturalists have . . . called into question the foundational categories that establish the canons of great works, the 'high' and 'low' culture divide, and the allegedly 'objective' scholarship that marks the exclusions within and between various disciplines" ("Racial Politics" 493).

29. I bracket "masculinist" to indicate that not all critical multiculturalists include feminist critiques. While some critical multiculturalists do expose the masculinist, phallocentric, sexist dimensions of this supremacist framework, others do not.

30. For example, Bonnie TuSmith maintains that,

> Given the long overdue inclusion of ethnic and women's writings into the curriculum, our understanding of American literary tradition—at least the version promoted in the 1950s and 1960s—must be revised. . . . At a minimum, we must distinguish . . . Eurocentric culture or literary tradition from ethnic traditions in this country. Since many of us agree that cultural diversity is an undeniable characteristic of American society, we must beware of applying monolithic terms to non-WASPS. Many times these terms simply do not fit. (21)

John Maitino and David Peck issue a similar call in their introduction to *Teaching Ethnic American Literature* where they explain that "traditional academic literary criticism has not worked as well with ethnic writers. Older criticism functions best with literary works that are finished, completed. Ethnic American literature is itself a process—in its stories of assimilation and resistance, of immigration and oppression—and demands a criticism that is equally flexible and fluid" (4).

31. Reginald Leamons Robinson offers an extensive discussion and analysis of this co-creative perspective. See especially his "Race, Myth and Narrative in the Social Construction of the Black Self." This type of co-creation is not identical with humanist notions of self-identity. For more on this issue, see Becky Francis' "Differences AND Commonalities."

32. For a critique of non-progressive forms of self-introspection, see Sarita Srivastava's " 'You're Calling Me a Racist?' The Moral and Emotional Regulation of Antiracism and Feminism." According to Srivastava, some 'white'-raced feminists, when struggling with their racism, become so deeply involved in self-reflection that they do not work for social change: "as some white feminists move toward new ideals of antiracist feminism, they often move toward deeper self-reflection rather than toward organizational change" (31).

33. Ella Shohat and Robert Stam make a similar point about what they call "polycentric multiculturalism": "all acts of verbal or cultural exchanges . . . tak[e] place not between discrete bounded individuals or cultures but rather between permeable, changing individuals and communities. Within an ongoing struggle of hegemony and resistance, each act of cultural interlocution leaves both interlocutors changed" (49). Unfortunately, they do not explore polycentric multiculturalism's potential for classroom teaching. See also Powell's assertion that "cultures do not exist in isolation but are inextricably intertwined in infinitely complicated ways. To study African American or Asian American or Chicano literature in isolation risks unwittingly reifying an essentialist conception of cultural identity" (*Ruthless Democracy* 10).

34. Thanks to Maria Elena Cepedes for introducing me to Fernando Ortiz's work.

35. Personal communication with the author.

36. In an earlier version of this essay, I took a more controlling view, and suggested that transformation "must be carefully strategized and designed." However, as Gloria Anzaldúa insightfully remarked when she read it, this desire for almost total control can make transformation less likely: "Transformation is not just something you can do with/through the mind—in fact strategizing and designing change means controlling it, which often defeats the process of change because control relies on the known, the old ways and what you want is the unknown, the new" (personal communication).

37. I coined the term *holistic-critical thinking* to counter my students' reactions to my emphasis on critical thinking. As I explain in more detail in chapter five, holistic-critical thinking is premised on interconnectivity and combines independent thought with a relational perspective.

38. "Commonalities" and "sameness" are not synonymous! As I explain in chapter three, commonalities, as I use the term, neither overlook nor deny the differences among us.

39. Throughout *Teaching Transformation*, I use the term 'whiteness' to indicate a framework, an epistemology and ethics, that functions as an invisible norm which undergirds U.S. culture (educational systems, the media, etc.). I do not automatically and always equate 'whiteness' with people identified as 'white'. Rather, the relationship between 'white' people and 'whiteness' is contingent (Frye, "White Woman Feminist"). I posit that we are all, in various ways, inscripted into this 'white'/supremacist framework. Because this framework functions to benefit 'white'-raced people, they can be more invested in it. This 'whiteness' intersects with certain versions of masculinity, economic status (middle- to upper-class), and colonization that has its roots in the Enlightenment.

40. U.S. people of color have studied 'whiteness' and racialized-'white' people for centuries. (See, for instance, William Apess's "Mirror;" David Walker's *Appeal, In Four Articles: Together with A Preamble to The Coloured Citizens of the World, but in Particular, and Very Expressly, to Those of the United States Of America*; John Gwaltney's *Drylongso*; *Ebony's The White Problem*; Mia Bay's *The White Image in the Black Mind: African-American Idea about White People, 1830–1925;* James Baldwin's *The Price of the Ticket*; Langston Hughes' *Ways of White Folks*; and the writings collected in David Roediger's *Black on White: Black Writers on What It Means to Be White*.) However, 'whiteness' studies—as an academic an academic field of study—is fairly new. It could be said to have been initiated in the early 1900s by W.E.B. Du Bois; since the 1990s, it has developed rapidly and spread widely. For overviews of the various forms academic 'whiteness'

studies takes, see Shelley Fisher Fishkin's "Interrogating 'Whiteness,' Complicating 'Blackness': Remapping American Culture"; Woody Doane's "Rethinking Whiteness Studies;" and Alexander Nguyen's "Souls of White Folk." As Christina Pruett notes, the fact that this focus on 'whiteness' "did not begin to reorient American scholarship until the last decade of the twentieth century suggests glaring disparities in cultural production that continue to authorize the long-standing practice and investment in racialist discourses of domination." As I use the term, 'whiteness' does not refer exclusively or even entirely to those people racialized as 'white'.

41. I borrow the term "of colors" from Indigo Violet and use it to underscore the diversity within racialized non-'white' groups; see her essay, "Linkages: A Personal-Political Journey with Feminist of Color Politics."

Chapter 1: "New" Stories for Social Change

1. I borrow this assertion from Inés Hernández-Ávila's "In the Presence of Spirit(s): A Meditation on the Politics of Solidarity and Transformation" 532.

2. Elana Dykwomon, "The Body Politic—Mediations on Identity" 454.

3. Gregory Cajete, *Native Science: Natural Laws of Interdependence* 105.

4. For more on core beliefs see Reginald Robinson. As he explains in "Human Agency, Subjectivity Negated," "A core belief flows from feelings and imaginations, and ordinary people reinforce this belief through words and deeds. From this core belief, ordinary people co-create their experiences and realities. Core beliefs, experiences, and realities are concentric circles, overlapping and indistinguishable" (1370).

5. Robinson explores the ways we co-create our racial reality in his essay, "Race Consciousness: A Mere Means of Preventing Escapes from the Control of Her White Masters?"

6. As David Lionel Smith points out, "[b]lack people can have white skin, blue eyes, and naturally straight hair; they can be half, three-quarters, seven-eighths or more white; they can even deny or not know that they are black. Claim what they will or look as they may, they are still by law and custom black" (180). For a philosophical exploration of this issue, see Naomi Zack's *Race and Mixed Race*.

7. I discuss personal agency and relational individualisms in "Transcendentalism Then and Now: Towards a Dialogic Theory and Praxis of U.S. Literature."

8. As john a. powell explains, individualism and 'whiteness' originated simultaneously and relationally:

> The ideology of individuality had its origins in the Enlightenment, which came concurrently with the emergence of Colonialism. During this germinative period, the essence of individualism was that Europeans were individuals as opposed to other people who were a "collective." The collectivity of the other served as a rationale and justification for the exploitation of the collective other. In other words, part of the longing to be a member of the dominant group was tied up with being an individual. In that sense, individuality was already racialized. Individuality and membership in the dominant culture meant something in particular in a specific moment related to white Europeans, although it was not clear at that moment that they were white. In fact, they were still in the process of becoming white. The ideology of individualism as opposed to the ideology of collectivity was part of the whiteness process. ("Disrupting Individualism" 3–4)

For further discussions of the colonialism, emergent 'white' supremacism, and other forms of elitism behind individualism's development, see powell's "Disrupting Individualism and Distributive Remedies with Intersubjectivity and Empowerment" and "Our Private Obsession, Our Public Sin," Joyce Warren's *The American Narcissus*, and David Goldberg's *Racist Culture*.

9. I use first person here because for many years I was enamored with this story of rugged individualism and tried to live my life according to its plot.

10. See also john a. powell's assertion: "How does individualism operate on race and gender today? The answer from the individualist camp is that race and gender do not exist. We are all just individuals. Any characteristic that does not reflect that individuality is simply an accident. In other words, any marker of gender, race, or sexuality around which meaning is constructed socially is largely irrelevant" ("Disrupting Individualism" 9).

11. In my experience, undergraduate students—especially those in their late teens and early twenties— have been extremely enmeshed in the story of self-enclosed individualism. See also Marc D. Rich and Aaron Castelan Cargile's discussion of the ways their students have been shaped by stories of self-enclosed individualism.

12. By "white-identified" I mean people who act and/or think in 'white' ways. Generally, but definitely not always, these people would be classified as 'white.' By "male-identified," I mean people who act and/or think in masculinist ways. Generally, but not always, these people would be classified as men.

13. See Thomas Friedman's *The World Is Flat* for an extensive discussion of this interconnectivity. Thanks to Jo-Ann Stankus for reminding me about this passage in Friedman's work. For another example of our economic interrelatedness, consider the September 11, 2001, bombings and their aftermath, the chain-reaction effect on the airline industry, airport businesses and related services, travel agencies, tourism, and so forth both in the United States and abroad. As Benjamin Barber notes, for many U.S. Americans, "September 11 was a brutal and perverse lesson in the inevitability of interdependence in the modern world" (26).

14. See also Gregory Cajete's discussion of our interconnectedness with the environment. As Cajete points out, "humans and the world interpenetrate one another at many levels, including the air we breathe, the carbon dioxide we contribute to the food we transform, and the chemical energy we transmute at every moment of our lives from birth to death" (*Native Science* 25).

15. Or, as Cajete states, "Language is more than a code; it is a way of participating with each other and the natural world. At one level language is a symbolic code for representing the world that we perceive with our senses. At the deeper psychological level, language is sensuous, evocative, filled with emotion, meaning, and spirit. Meanings are not solely connected to intellectual definition but to the life of the body and the spirit of the speaker. In its holistic and natural sense, language is animate and animating, it expresses our living spirit through sound and the emotion with which we speak. In the Native perspective, language exemplifies our communion with nature rather than our separation from it" (*Native Science* 72).

16. Thich Nhat Hahn has coined the phrase "inter-being" to describe this interconnectedness; see his discussions in *Interbeing: Fourteen Guidelines for Engaged Buddhism* and *Transformation at the Base.* For additional discussions of Nhat Hanh's concept see Chân Không's *Learning True Love: How I Learned and Practiced Social Change in Vietnam* and Robert H. King's *Thomas Merton and Thich Nhat Hahn* (102–03).

17. See also James Cone's assertion that "Martin Luther King was right: We are bound together as one humanity. What affects one directly affects all indirectly. The sooner we realize the interconnectedness of our social existence, the sooner we will end our isolation from each other and begin to develop ways that we can work together toward the liberation of the poor throughout the world" ("Afterword" 200). The Dalai Lama makes a similar point: "our individual well-being is intimately connected both with that of all others and with the environment within which we live. It also becomes apparent that our every action, our every deed, word, and thought, no matter how slight or inconsequential it may seem, has an implication not only for ourselves but for all others, too" (41).

18. For an illustration of this type of mutual encounter, see Anzaldúa's discussion of nepantleras in the final two sections of her essay, "now let us shift."

19. Macy also notes that the larger wholes are distinct creations with unique capabilities: "At each holonic level new properties and new possibilities emerge, which could not have been predicted.

From the respective qualities of oxygen and hydrogen, for example, you would never guess the properties that emerge when they interact to make water."

20. bell hooks makes a related point: "All that we do to break away from the idea of separate ego and to acknowledge our interdependence is already a radical step away from race, nationality, religious affiliation, sexual preference, class positionality, and educational status as fixed markers in life" ("Buddhism and the Politics of Domination" 60–61).

21. I am grateful to Christa Downer who pointed out that my pedagogy is invitational, rather than persuasive. See her discussion in "The Making of a Pluralistic Egalitarian Society: Reconceptualizing the Rhetoric of Multiculturalism for the 21st Century."

22. *Desconocimientos* is a term Anzaldúa coined to describe individual and collective ignorances. Sometimes, desconocimientos are intentional (e.g., a willed resistance to consciously acknowledging a painful truth about one's self or one's culture), but other times they can be unintentional (though dangerous and destructive) limitations in one's education and upbringing. For more on desconocimientos see her essay "now let us shift," especially section 3 "*the Coatlicue state . . . desconocimiento and the cost of knowing*" (550–54).

23. I am grateful to Mukamtagara Jendayi for asking me to clarify the meaning of this statement in my graduate course WS 5663, Transgressive Identities (Summer 2006).

24. For an example of how this type of guilt can shift the classroom focus, see Anzaldúa's "haciendo caras, una entrada" xix–xx.

25. To underscore this point, I have recently begun including a discussion of listening, or what I call "listening with raw openness," on my syllabi. (See appendix 2.)

26. I first learned about discussion questions from Karen Anderson (University of Arizona) who suggested a format using notecards collected at the beginning of each class period. I've modified the format in several ways, especially at the graduate course level.

27. See the syllabi in appendix 6 for examples of how I use discussion questions.

28. Like Anzaldúa, I "frequently check [my] understanding of the other's meaning, responding with, 'Yes, I hear you. Let me repeat your words to make sure I'm reading you right' " ("now let us shift" 569–70).

29. As Arthur Chickering notes, "One major finding, consistent with a substantial body of psychological research, is that the brain does not simply reproduce external reality. Instead, apparently, about 80 percent of what we perceive and think we have understood is in fact rooted in prior attitudes, information, ideas, and emotional reflexes" ("Curricular Content" 131). For more on this topic, see also Reginald Robinson's distinction between "mind concepts" and reality ("Human Agency, Negate Subjectivity" 1369).

30. See also Linda James Myer's discussion of language. As Myers explains, "We must recognize the power of words, for words label experience and thus give form to thoughts" (26). Joy Harjo makes a similar point: "Language is culture, a resonant life form itself that acts on the people and the people on it. The worldview, values, relationships of all kinds—everything, in fact, is addressed in and through a language" (*Spiral of Memory* 99). See also Kelly Oliver: "What people around us believe and feel about the world and others also influences what we see" (*Witnessing* 14).

31. Anzaldúa makes a similar point in "now let us shift" as she explores the ways identity and culture shape our perceptions: "Your identity is a filtering screen limiting your awareness to a fraction of your reality. What you or your cultures believe to be true is provisional and depends on a specific perspective. What your eyes, ears, and other physical senses perceive is not the whole picture but one determined by your core beliefs and prevailing societal assumptions" (542).

32. For discussions of the ways these concepts damage us, see Linda Myers et al. "Identity Development and Worldview: Toward an Optimal Conceptualization." As the authors explain,

> a positive self-identity is not easily attained in a culture such as this one. This premise is supported by the pervasive number of "-isms" (racism, sexism and ageism) and their impact

on those who would be defined as inferior by the dominant way of perceiving in this society. . . . [T]he very nature of the conceptual system is itself inherently oppressive and . . . all who adhere to it will have a difficult time developing and maintaining a positive identity. (55)

See also Reginald Robinson's "Race Consciousness."

Chapter 2: Forging Commonalities

1. Marilou Awiatka, *Selu: Seeking the Corn-Mother's Wisdom*, 155.
2. See also Jace Weaver's introductory comments in *That the People Might Live*.
3. For more information on this ongoing history of conquest, see the essays in M. Annette Jaimes's edited collection on *The State of Native America: Genocide, Colonization, and Resistance*.
4. As I explained in the previous chapter, *desconocimientos* is Gloria Anzaldúa's term for individual and collective ignorances.
5. Some of the texts I've used are Benjamin Franklin's *Autobiography*, Frederick Douglass's *Narrative*, Harriet Jacob's *Incidents*, Horatio Alger's *Ragged Dick*, Amy Tan's *The Joy Luck Club*, Anzaldúa's *Borderlands/La Frontera*, and Sandra Cisnero's *House on Mango Street*.
6. I use "we" intentionally to suggest that through this discussion, the students become aware of their own presuppositions.
7. As Wayne Charles Miller explains, the Puritans were distinctly influenced by the Bible:

   they possessed a book, the Bible, which, while forming the basis of their faith, also contained stories and a history which provided them with a world view, with a distinct definition of man's [sic] role in nature as defined by the Puritans, and with a repository of metaphor and analogy that they, and other white Europeans as well, would use as a means to transform the continent—to re-define it within the frame of their own cultural consciousness, particularly so in regard to their definition of man's role in nature, a definition that represented a radical departure from the world view of the red men [sic] who had preceded them. Seen from this perspective, the various Christian theological discussions, the accounts of the establishment of European-based civilizations in the wilderness, and the narratives of travel and exploration—often laced with Biblical allusions—all become the written testimony of the efforts to establish their culture in a place strange to them. (31–32)

   For discussions of how Native peoples' worldviews were shaped by their creation stories, see Paula Gunn Allen's *The Sacred Hoop* and Gregory Cajete's *Native Science*.
8. See Paulo Freire for a discussion of the banking account of education.
9. For useful discussions of Native peoples' creation stories and the ways they influenced their worldviews and social systems, see Allen's *The Sacred Hoop*, chapter four, Vine DeLoria's *God Is Red*, especially chapter three, and Gregory Cajete's *Look to the Mountain*.
10. These authors are all included in the *Heath Anthology of American Literature*. For an example of some of the ways I have grouped texts together, see my sample syllabi in appendix 6.
11. As Sacvan Bercovitch notes, this ideology—which he defines as "individualism, progress, and the American Way"—has lead to an unspoken set of guidelines for living and a "system of values so deeply ingrained" in the national imaginary that it seems entirely natural (420).
12. See also Thomas King's discussion of Christian and Indigenous creation myths, where he makes a related point:

    In Genesis all creative power is vested in a single deity who is omniscient, omnipotent, and omnipresent. The universe begins with his thought, and it is through his actions that it comes into being. In the Earth Diver story, and in many other Native creation stories for that matter, deities are generally figures of limited power and persuasion, and the acts of creation and the decisions that affect the world are shred with other characters in the drama. (24)

13. For critiques of this appropriation, see Kathryn Shanley's "The Indians America Loves to Love and Read: American Indian Identity and Cultural Appropriation;" Lisa Aldred's "Plastic Shamans and Astroturf Sun Dances;" Cynthia A. Snaveley's "Native American Spirituality;" and Michael York's "New Age Commodification and Appropriation of Spirituality."
14. Not all Judeo-Christian worldviews are hierarchical and dualistic. For a holistic approach, see Matthew Fox.
15. See chapter one for a discussion of status-quo stories.
16. See, for example, McLaren's "Multiculturalism and the Postmodern Critique," McLaren and Tomaz Tadeu da Silva's "Decentering Pedagogy," and Richard Delgado's *The Rodrigo Chronicles*.

Chapter 3: (De)Constructing 'Race'

1. Kwame Anthony Appiah, "The Conservation of 'Race' " 43.
2. john a. powell, "A Minority-Majority Nation: Racing the Population in the Twenty-First Century" 1403.
3. Reginald Leamon Robinson, " 'Expert' Knowledge: Introductory Comments on Race Consciousness" 150.
4. See also powell's assertion, "Oppositional efforts to valorize what has been denigrated by the dominant society are an attempt to challenge the racing or Othering process. Attempts have consisted of claiming a voice as a subject in opposition to the dominant discourse. But *voice is often trapped within the unexamined language and symbols of the dominant group that it wishes to reject*" ("Whites Will Be Whites" 433, my emphasis).
5. See also Reginald Leamon Robinson's assertion that "race exists, if ever, in our individual and cultural consciousness. If we do not constantly and consciously meditate on it, race cannot exist. Unfortunately, we fuel this social construct with our mental kindling and intellectual logs" ("Shifting" 234).
6. Crispin Sartwell makes a similar point:

> What is "normal" in our culture is always, invisibly, raced white. . . . The opponents of multiculturalism often appeal to the hackneyed image of the melting-pot: they ask why we can't all form a single people, and they offer to welcome African-Americans and other minorities into European-American culture with apparently open arms. (As one white student in my African-American philosophy class put it, "Can't we just forget about race? Let's just all be the same.") But this argument rests on two crucial and intertwined errors. First, it equates American culture with European-American culture, and it offers this culture to everyone in the form of the canon of Western literature, science, and philosophy. . . . Thus, second, the "neutral" culture which these folks offer minorities is in fact raced white. (9–10)

7. I am referring here to a specific group of students back in the early 1990s. These students held conservative racial ideas; for instance, students whom I would have described as "mexicano" or "Chicana" consistently self-identified as "Spanish" (read: 'white'). Given the more recent changes in racialized U.S. discourse, fewer students today would be as stunned by the invitation to consider 'whites' as a 'race.' However, often today's students (especially but not only those of European descent) become resentful when invited to explore 'whiteness'. For a discussion of these changes in students' racialized perceptions, see Charles Gallagher's "White Reconstruction in the University."

8. For an excellent discussion of how students tend to think in terms of stereotyped binary oppositions see Sharon Stockton's "Blacks vs. Browns."

9. Early influential texts in ethnic studies and critical 'race' theory include Michael Omi and Harold Winant's *Racial Formation in the United States from the 1960s to the 1980s*; Patricia Williams's *The Alchemy of Race and Rights: Diary of a Law Professor*; and Derrick Bell's *Faces at the Bottom of the Well*, especially the last chapter. For historical investigations of 'white' identities, see Theodore W. Allen's *The Invention of the White Race*, Parts I and II; Noel Ignatiev's *How the Irish Became White*; Karen Brodkin's *How Jews Became White Folks and What that Says about Race in America*; and Ian Haney-López's *White By Law: The Legal Construction of Race*. For investigations of contemporary 'white'-raced people, see John Hartigan's *Racial Situations: Class Predicaments of Whiteness in Detroit*; Ruth Frankenberg's *The Social Construction of Whiteness: White Women, Race Matters*; Pamela Perry's "White Means Never Having to Say You're Ethnic;" and the essays collected in Ashley W. Doane and Eduardo Bonilla-Silva's *White Out: The Continuing Significance of Racism*. For groundbreaking analyses of 'whiteness' in cultural texts see especially Richard Dyer's "White" and Toni Morrison's *Playing in the Dark*. See also Shelley Fisher Fishkin's "Interrogating 'Whiteness,' Complicating 'Blackness': Remapping American Culture;" Valerie Babb's *Whiteness Visible: The Meaning of Whiteness in American Literature and Culture*; Fred Pfeil's *White Guys: Studies in Postmodern Dominance and Difference*; and Owen J. Dwyer and John Paul Jones III's "White Socio-Spatial Epistemology."

10. For more on the potential risks involved in 'whiteness' studies, see Andersen's "Whitewashing Race;" Mike Hill's discussion of "academic great white hype" (*After Whiteness* 3); and Ann E. Kaplan's observation that "focusing on whites' re-perception of themselves in the category of white, though a new perspective, may produce a different kind of obsession with whiteness that is nevertheless still an obsession" (324).

11. As Joe Kincheloe notes in "The Struggle to Define and Reinvent Whiteness," "the concept [of 'whiteness'] is slippery and elusive" (162).

12. Morrison makes a similar point in her analysis of canonical U.S. literature when she maintains that this unacknowledged 'whiteness' has created a literary "language that can powerfully evoke and enforce hidden signs of racial superiority, cultural hegemony, and dismissive 'othering' " (*Playing* x–xi). But see also Giroux's "Rewriting the Discourse of Racial Identity" where Giroux argues that, during the 1990s, "Whiteness became increasingly visible as a symbol of racial identity. Displaced from its widely understood status as an unnamed, universal moral referent, Whiteness as a category of racial identity was appropriated by diverse conservative and right-wing groups, as well as critical scholars, as part of a broader articulation of race and difference."

13. Of course, this notion of "differences based on blood" is itself the product of pseudo-scientific, racist thinking. See Zack's *Race and Mixed Race* 13.

14. For more on the ways 'whiteness' has been used to stigmatize those labeled non-'white', see Ross Chambers's discussion of what he calls "examinability." As Chambers explains, "Examinability, as I am using the term here, refers to the unfavorable attention to certain groups that is recognized, in its overtly hostile forms, by terms like *misogyny, homophobia*, and *racism* but that has many other more covert, polite, hypocritical, and even sanctimonious forms" (194, his emphasis).

15. Frye's method here is innovative, given that she was writing in the early 1990s when few other 'white'-identified scholars had investigated 'white' racial issues. She builds her analysis from a variety of perspectives, including some by 'white'-identified people. She draws from John Gwaltney's *Drylongso: A Self-Portrait of Black America*; Cherríe Moraga and Gloria Anzaldúa's edited collection, *This Bridge Called My Back: Writings by Radical Women of Color* (specific texts are those by Barbara Cameron, Chrystos, doris davenport, and Mitsuye Yamada); bell hooks's *Feminist Theory: From Margin to Center*; her own experiences; and the words of Minnie Bruce Pratt, a 'white'-identified author. See also Frye's earlier essay, "On Being White: Toward a Feminist Understanding of Race and Race Supremacy."

16. As George Yancy asserts, "One can cease to cooperate with structures of white power; cease to perform white racist acts; and, hence, help to dismantle structures of white power. Whites must come to see how they have become seduced by whiteness, and how they make choices based upon that seduction" ("Introduction" 8). For another discussion of the contingent nature of 'whiteness' and those people racialized as 'white,' see Zeus Leonardo's "The Souls of White Folk: Critical Pedagogy, Whiteness Studies, and Globalization Discourse."

17. For discussions of the shift to defining Mexican Americans and other "Latinos" as an ethnicity, see Ian Haney-López's "Race on the 2010 Census: Hispanics and the Shrinking White Majority" and Victoria Hattam's "Ethnicity and the Boundaries of Race: Rereading Directive 15."

18. Hattam also notes that in this 2000 census more Latinos than in any other census refused to self-identify as a designate 'race': "[a]n unprecedented 42.2 percent of Hispanics checked 'some other race' on the 2000 census" (66).

19. See for instance Hattam's "Ethnicity and the Boundaries of Race" and Patricia Palacio Paredes's "Note: Latinos and the Census: Responding to the Race Question."

20. See also Winthrop Jordan's discussion of the early colonists in *White Man's Burden*. As Jordan explains, "From the initially most common term *Christian*, at mid-century there was a marked shift toward the terms *English* and *Free*. After about 1800, taking the colonies as a whole, a new term of self-identification appeared–*white*" (52, his italics).

21. Zackodnik goes on to explain that the colonists' "initial focus on spirit arguably recurs in a later focus on morality and moral legitimacy as the mulatto's 'predisposed' acts and behaviors become a focus and already open the door to race as contingent upon the social" (424). For another discussion of the ways religious and racial boundaries overlap, see john a. powell's "Whiteness: Some Critical Perspectives: Dreaming of a Self Beyond Whiteness and Isolation." As powell explains, "The precursor of [North American] racial boundaries was a religious boundary, Christian and non-Christian. In some ways, the first 'racial benefit' in the United States was that a Christian could not be a permanent slave" (20). For additional histories of the racialization of 'white' people, see Theodore Allen's *Invention of the White Race*, volume I; Thandeka's *Learning to Be White*; Alastair Bonnett's "Constructions of Whiteness;" David Roediger's *Toward the Abolition of Whiteness*; and Noel Ignatiev's *How the Irish Became White*.

22. For additional analyses of how slavery was racialized, see Zackodnik's "Fixing the Color Line," Thandeka's *Learning to Be White*, and Collette Guillaumin's "Race and Nature."

23. The lyrics, from William "Smokey" Robinson's "The Black American," continue:

> Your heritage is right here now, no matter what you call yourself or what you say And a lot of people died to make it that way. . . . God knows we've earned the right to be called American Americans and be free at last. And rather than you movin' forward progress, you dwelling in the past. We've struggled too long; we've come too far. Instead of focusing on who we were, let's be proud of who we are. We are the only people whose name is always a trend. When is this shit gonna end? (Qtd. in Norwood 146)

24. For discussions of the shifting definitions of 'white' people see Valerie Babb's *Whiteness Visible*; Matthew Jacobson's *Whiteness of a Different Color: European Immigrants and the Alchemy of Race*; Noel Ignatiev's *How the Irish Became White*; Elizabeth Grace Hale's *Making Whiteness*; Omi and Winant's *Racial Formation in the United States from the 1960s to the 1980s*; and john a. powell's "Our Private Obsession."

25. For additional discussions of this issue, see Joe L Kincheloe and Shirley R. Steinberg's "Addressing the Crisis of Whiteness."

26. Orlando Patterson makes a related point, which he applies to "liberal intellectuals" as a whole: "Having demolished and condemned as racist the idea that observed group differences have any objective, biological foundation, the liberal intellectual community has revived the 'race' concept as an essential category of human experience with as much ontological validity as the discarded racist notion of biologically distinct groups" (72). See his argument throughout chapter 1 of

The Ordeal of Integration. For a discussion of social scientists' reification of 'race', see Mica Pollock's "Wrestling with Race."

27. Unfortunately, however, because Morrison does not move outside of 'black'/'white' binaries, she overlooks and even dismisses the ways other racialized identities have shaped literary 'whiteness.' (See especially *Playing in the Dark* 48.)

28. It would also be very helpful and illuminating if more scholars and theorists would at least occasionally consider the possible impact their theories might have on classroom teaching.

29. Shohat and Stam make a similar point, arguing that they "would therefore distinguish between a personalistic, neurotic guilt on the one hand, and a sense of collective and reciprocal answerability on the other" (343–44).

30. I discuss my use of racialized labels at greater length in chapter four.

31. I also use Zeus Leonardo's description of 'whiteness' as "a collection of everyday strategies" (32).

32. For an extensive discussion of the ways 'white'-raced people have benefited from this framework, see George Lipsitz's *The Possessive Investment in Whiteness: How White People Profit from Identity Politics*.

33. Dwyer and Jones go on to note that this 'white' "spatial epistemology . . . also relies upon the ability to survey and navigate social space from a position of authority" (210). See also Dwyer and Jones's discussion of the

> categorical naturalization . . . [which] underwrites private property and the construction and orderly maintenance of segmented social spaces, from gated communities to redlined districts, from nature "preserves" . . . to office towers. . . . Simultaneously marking and making difference by bounding white and Other in their respective places, this racialized geography has been reproduced on and through the built environment throughout American history. (212–13, my italics)

34. See Lipsitz's book, *The Possessive Investment of Whiteness*.

Chapter 4: Reading 'Whiteness,' Unreading 'Race'

1. For other theorists demanding an interrogation of 'whiteness' in pedagogy, see Chandra Mohanty's "On Race and Voice," bell hooks' *Talking Back*, and Henry Giroux's "Racial Politics, Pedagogy, and the Crisis of Representation in Academic Multiculturalism" and "Rewriting the Discourse of Racial Identity: Towards a Pedagogy and Politics of Whiteness."

2. For useful analyses of this 'white' crisis and the development of reactionary 'white'-raced identities, see Giroux's "Rewriting the Discourse of Racial Identity: Towards a Pedagogy and Politics of Whiteness," Joe L. Kincheloe and Shirley R. Steinberg's "Addressing the Crisis of Whiteness," Charles Gallagher's "White Construction in the University," and Peter McLaren's *Revolutionary Multiculturalism*. According to McLaren, "Feeling that their status is now under siege, whites are now constructing their identities in reaction to what they feel to be the 'politically correct' challenge to white privilege" (262).

3. Henry Giroux makes a similar point: "The issue of making White students responsive to the politics of racial privilege is fraught with the fear and anger that accompany having to rethink one's identity. Engaging in forms of teaching that prompt White students to examine their social practices and belief systems in racial terms may work to reinforce the safe assumption that race is a stable category, a biological given, rather than a historical and cultural construction" ("Rewriting" 313).

4. Joe Kincheloe and Shirley Steinberg make this argument in their article, "Addressing the Crisis of Whiteness" (12). See also Kincheloe's discussion in "The Struggle to Define and Reinvent Whiteness."

5. As Margaret Andersen notes,

> Joe Kincheloe and Shirley Steinberg state that a key goal of whiteness studies is "creating a positive, proud, attractive, antiracist white identity that is empowered to travel in and out of various racial/ethnic circles with confidence and empathy" (1998:12). Who gets the privilege of such "traveling"—that is, assuming the stance of the outsider with the option of leaving at any time? Who gets to be a "race" without the negative consequences of racial stratification? (32)

6. George Yancy makes a similar point, describing 'whiteness' as "a form of inheritance" which one can choose "not [to] accept" ("Introduction" 8).

7. See also Toni Morrison's assertion in *Playing in the Dark*: "until very recently, and regardless of the race of the author, the readers of virtually all of American fiction have been positioned as white" (xii).

8. Octavia Butler is the author of twelve novels and a collection of short stories. Winner of science fiction's two highest honors, the Hugo and the Nebula, Butler was the first science fiction writer (of any color or gender) to receive the MacArthur Foundation "Genius" Grant. As one of the first African-American science fiction writers, her work represents a significant breakthrough. Her novels contain strong female protagonists whose wisdom and actions make them agents of change. She deals with complex sophisticated issues, including the struggle for power and control; the ways these struggles are inflected by gender, ethnicity/'race,' and class; fear of and confrontation with differences; and the creation of new communities where peoples of many colors, and often different species, interact.

9. The only protagonist not racialized is Teray in *Patternmaster*.

10. Earlier in the narrative Butler informs us that Lilith's married name is "Iyapo" and that her husband was originally from Nigeria; however, these cues do not necessarily tell readers anything about Lilith's own ethnicity.

11. Thus, for example, Donna Haraway states that "[i]llustrating the workings of the unmarked category, 'white,' *Dawn*'s cover art has allowed several readers whom I know to read the book without noticing either the textual cues indicating that Lilith is black or the multi-racialism pervading Xenogenesis" (381). She is referring to the novel's first edition, which has a racially ambiguous cover.

12. The Afrocentric nature of Lauren's surname (Olamina) does provide an important clue to her ethnicity for readers aware of the history of African-American identities.

13. It is significant (but not at all surprising) that in this dystopic future, where violence and intense disrespect for human life reign, people self-segregate based on color.

14. Earlier, we learn that Kevin has "the kind of pale, almost colorless eyes that made him seem distant and angry whether he was or not" (*Kindred* 13). But since some "black"-identified people have pale eyes, this brief description generally does not inform readers of Kevin's color. In my experience teaching this novel, it certainly does not prepare readers for the shock they experience when they discover that Kevin is not "black."

15. Toni Morrison employs a very different strategy and yet attains remarkably similar effects in *Paradise*. The novel opens with the sentence "They shoot the white woman first." However, even after reading the entire novel, readers cannot know which woman is 'white.' This ambiguity is intentional. As Morrison asserts in a 1998 interview, "My point was to *flag* race and then to erase it, and to have the reader believe, finally—after you know *everything* about these women—their interior lives, their pasts, their behavior—that the one piece of information you don't know, which is the race, may not in fact matter. And when you *do* know it, what do you know? [smiles]" (Interview, her emphasis).

16. Unfortunately, Shipler only focuses on 'white'/'black' interactions; however, his remarks can be extended to other racialized groups as well.

17. Unless, of course, they have looked carefully at the more recent book cover and read the author's bio. I have been struck by how few of my students (especially at the undergraduate level) pay attention to either.

18. As I argued in chapter 3, popular beliefs to the contrary, there are no genetically distinct 'races' of people. 'Race' is, rather, an economically and politically motivated classification system with highly destructive effects; and racialized identities are unstable, artificial, and relational. For discussions of the roles nineteenth-century pseudo-science played in naturalizing racialized taxonomies, see Kwame Anthony Appiah's "The Uncompleted Argument: Du Bois and the Illusion of Race" and Naomi Zack's *Race and Mixed Race*.

19. In keeping with my (de)racializing approach, I did not describe this unit to the students in my Introduction to Literature course *as* a unit on 'whiteness.'

20. I explain that southern Europeans, light-skinned Jews, Irish, and Catholics of European descent, for example—were most definitely *not* 'white' in eighteenth- and nineteenth-century America. See Yehudi Webster's *The Racialization of America* (132–33); Ignatiev's *How the Irish Became White*, and Theodore Allen's *The Invention of the White Race*.

21. See Thandeka's discussion of this resistance in *Learning to Be White* (47–51).

22. "It was estimated on the basis of intercensual and birth-date rate comparisons that 25,000 blacks 'passed into the general community' *each year* from 1900–10" (Goldberg 344, n. 12, his italics). See also Linda Alcoff, David Lionel Smith, and Naomi Zack.

23. There are too many examples of this selective racialization to list, but for a few additional examples see DeLillo's references to the "black" Pentecostal (135), the "Asian" child (154), the "Indians" (or are they "Pakistanis"? Jack's family doesn't know for sure) at drugstores (179), and the "Iranian" who delivers the newspaper (184).

24. See also Rebecca Aanerud's assertion that "[r]eading whiteness into texts . . . that are not overtly about race is an essential step toward disrupting whiteness as the unchallenged racial norm" (43).

Chapter 5: Teaching the Other?

1. I use the terms *trans* and *transpeople* rather than *transgender* or *transsexual* in order to be more inclusionary. For discussions of terminology, see Jamison Green's *Becoming a Visible Man*, Riki Wilchin's *Read My Lips*, and Max Wolf Valerio's " 'Now That You're a White Man': Changing Sex in a Postmodern World—Being, Becoming, and Borders." I went back and forth on whether to include trans issues in this chapter. Because they have more to do with gender than with sexuality, to include trans issues here risks reinforcing stereotypical views of transpeople. Moreover, scholars exoticize trans issues/people or use them simply to further their own careers. As Riki Wilchins asserts,

> Academics, freaks, and feminist theorists have traveled through our lives and problems like tourists on a junket. Picnicking on our identities like flies at a free lunch, they have selected the tastiest tidbits with which to illustrate a theory or push a book. The fact that we are a community under fire, a people at risk, is irrelevant to them . . . It is not intended to help, but rather to explicate us as Today's Special: trans under glad, or perhaps only gender a la mode. . . . No one bothers to investigate the actual conditions of our lives or the lives of those we hold dear. No one asks about the crushing loneliness of so many translives, or about sexual dysfunction. (22)

I do not want to travel through trans lives in these ways. However, after much thought and many drafts, I decided that I had to include trans issues both because I include them in my teaching and because they are so often overlooked. To exclude them here would participate in this erasure and dishonor my ongoing efforts to enact a radically inclusionary politics. Moreover, although trans issues and identities focus more closely on gender than on sexuality, they have sometimes been included in queer theory.

2. According to Foster,

> An equally irresponsible stance is to abandon all responsibility for understanding and inter-preting or to neglect to determine what relevant literary methodology there is for dealing with "Other" literatures because we've discovered that to Others we are also Other. Chagrined and apologetic, some of us murmur that "Since I'm not X, I can't possibly know." We resort to summarizing what a prominent "scholar" has said or to confiding our "feelings" about literary experience. While our personal feelings are important, sometimes even to the matter at hand, confessing is not professing. This stance betrays itself very simply, I think. How many of us are, or have students who are, nineteenth-century factory workers? How many of us are confused because our mother has married our father's murderer? Yet we know and expect others to appreciate *Life in the Iron Mills* and *Hamlet*. (199)

3. Not her real name.

4. For a more extensive discussion of multiculturalism see the introduction to this book.

5. I discuss the performative dimensions of language in chapter two and in my earlier book, *Women Reading Women Writing*. See also Cynthia Dillard's discussion of the Afrocentric dimensions of language in *On Spiritual Strivings*, especially 91–92.

6. I define *intellectual* broadly to include intuition, imagination, and emotional ways of knowing as well as rational thought and logical analysis.

7. For an excellent definition of the concept of holism, see Ramón Gallegos Nava. As he explains,

> The term "holistic" comes from the Greek holos, which in our context means wholeness. The term refers to comprehending reality as a function of a whole in integrated processes. The term "holistic" is used to denote that reality is an undivided whole; that it is not fragmented; that the entirety is the fundamental reality. The whole, from such a perspective, is not a static structure, but a universal impermanent flow. The type of intelligence necessary to comprehend it is dif-ferent from mechanical thought. . . . The holistic vision is based on a certainty that everything is interconnected. We are united with all others and with the non-human forms surrounding us in a complex network of life. (14)

8. See chapter one for a description of the discussion question assignments I build into my courses.

9. I intentionally listed these five identities (gay, lesbian, bisexual, homosexual, and queer) because I wanted to subtly challenge students' monolithic thinking and encourage them to recognize the diverse categories people use to define their sexuality.

10. Coming out as lesbian/gay/bisexual/queer is still a highly charged decision. As David Wallace notes, "acceptance of people politicized as queer—even in the relatively liberal confines of the academy—is far from complete" (65). For more on the various issues affecting lesbians in higher education, see the essays collected in Linda Garber's *Tilting the Tower: Lesbian/Teaching in the Queer Nineties*.

11. As Naomi Zack points out, "It has been estimated that between 70 and 80 percent of all designated black Americans have some degree of white ancestry" (*Race* 75). Adrian Piper makes a similar point concerning "white"-identified Americans in "Passing for White, Passing for Black." For discussions of hypodescent see Zack's "Introduction" to *American Mixed Race* and Maria Root's "The Multiracial Contribution to the Psychological Browning of America." For the (non)history of mixed-race peoples, see Zack's *Race and Mixed Race*.

12. John Alberti, who *does* identify as "white," employs a similar strategy, which he calls "racial distancing" (211).

13. For stories by Sui Sin Far, Jewett, Dunbar-Nelson, and Freeman, see Judith Fetterley and Marjorie Pryse's collection, *American Women Regionalists, 1850–1910*.

14. See, for example, Paula Bennett, Lillian Faderman, and Eve Sedgwick (*Tendencies*) on Dickinson; Michael Moon on Whitman; and Sedgwick (*Epistemology* and *Tendencies*) on James.

15. I follow bell hooks in using the phrase *feminist movement*, rather than *the feminist movement*. To my mind, the former term is less monolithic than the latter; also, it more fully captures both the

unity and the diversity among contemporary feminists. See her *Feminist Theory: From Margin to Center.*

16. For an excellent discussion of transchildren, see Jody Norton's "Transchildren, Changelings, and Fairies: Living the Dream and Surviving the Nightmare in Contemporary America."

17. There are many examples of these omissions. See, for example, Juan Bruce-Novoa's discussion of how John Rechy and Sheila Ortiz Taylor were shut out of Chicano literature canon formation as well as Barbara Smith's discussion of the erasure of black lesbian writing in "Towards a Black Feminist Criticism" and "The Truth that Never Hurts."

18. For an overview of various ways nonwestern cultures have viewed homosexuality, see David F. Greenburg's *The Construction of Homosexuality* and Walter Williams's *The Spirit and the Flesh.* In *Sexual Dissidence: Augustine to Wilde, Freud to Foucault* Jonathan Dollimore discusses some of the reasons many contemporary westerners see homosexuality as so threatening.

19. Suzanne Pharr's *Homophobia: A Weapon of Sexism* and the essays collected in Warren Blumenfeld's *Homophobia: How We All Pay the Price* provide thought-provoking discussions of these issues.

20. For Anzaldúa's assertion, see Keating, "Writing, Politics, and *las Lesberadas: Platicando con* Gloria Anzaldúa." For Anzaldúa's discussion of her sexual history, see her *Interviews/Entrevistas.*

21. See the narratives collected in Carol Queen and Laurence Schimel's *Pomosexuals.* The contributors describe a variety of transgressive experiences, ranging from gay men who have sex with women to self-identified lesbians who undergo sex-change surgery and become heterosexual men.

22. Diana Fuss's introduction to *Inside/Outside* and Teresa de Lauretis's "Queer Theory: Lesbian and Gay Sexualities, An Introduction" provide useful overviews of recent debates in lesbian/gay theory, especially when read in conjunction with "To(o) Queer the Writer," Gloria Anzaldúa's critique of academic queer theory.

23. Ginzberg and Phelan provide useful overviews of these debates on lesbian identities. See also Henry Rubin's insightful discussion of how the rise of lesbian feminism in the 1970s negatively impacted butch lesbians in chapter 2 of his *Self-Made Men: Identity and Embodiment among Transsexual Men.* He finds a significant correlation between the increased number of transmen and lesbian feminists' exclusionary gender views.

24. See, for example, Anzaldúa's "To(o) Queer the Writer," *Borderlands/La Frontera,* and *Interviews/Entrevistas.*

25. A number of books and essays explore the social construction of sexuality. Michel Foucault's groundbreaking *The History of Sexuality* examines the development of sexual identities in western culture, and Halperin's *One Hundred Years of Homosexuality* puts contemporary conceptions of sexuality into historical perspective by providing an overview of the various ways westerners have perceived gender and sexual identities. Donald Hall's *Queer Theories* offers a short discussion of the development of sexual identities that works well with students.

Conclusion: Transforming Status-Quo Stories

1. Patricia Williams, *Seeing a Color-Blind Future: The Paradox of Race* 16.
2. I borrow the term "see through" from Anzaldúa. See her discussion in "now let us shift."

REFERENCES

Aanerud, Rebecca. "Fictions of Whiteness: Speaking the Names of Whiteness in U.S. Literature." Frankenberg, *Displacing Whiteness: Essays in Social and Cultural Criticism* 35–59.

Alberti, John. "Teaching the Rhetoric of Race: A Rhetorical Approach to Multicultural Pedagogy." Brannon and Greene, *Rethinking American Literature* 203–15.

Alcoff, Linda. "Mestizo Identity." Zack, *American Mixed Race* 257–78.

Aldred, Lisa. "Plastic Shamans and Astroturf Sun Dances: New Age Commercialization of Native American Spirituality." *American Indian Quarterly* 24 (2000): 329–52.

Alexander, M. Jacqui. *Pedagogies of Crossing: Meditations on Feminism, Sexual Politics, Memory, and the Sacred.* Durham, NC: Duke UP, 2005.

———. "Remembering *This Bridge*, Remembering Ourselves: Yearning, Memory, and Desire." Anzaldúa and Keating, *this bridge we call home* 81–103.

Allen, Paula Gunn. *The Sacred Hoop: Recovering the Feminine in American Indian Traditions.* Boston: Beacon, 1986.

Allen, Theodore W. *The Invention of the White Race: Racial Oppression and Social Control.* Vol. I. London and New York: Verso, 1994.

———. *The Invention of the White Race: The Origin of Racial Oppression in Anglo-America.* Vol. II. London and New York: Verso, 1997.

Andersen, Margaret L. "Whitewashing Race." Doane and Bonilla-Silva, *White Out* 21–34.

Anzaldúa, Gloria E. "El Paisano Is a Bird of Good Omen." *Cuentos: Stories by Latinas.* Ed. Alma Gómez, Cherríe Moraga, and Mariana Romo-Carmona. Kitchen Table P, 1983. 153–75.

———. "Foreword to the Second Edition." Moraga and Anzaldúa, *This Bridge Called My Back.* N. pag.

———. "Geographies of the Self: Re-Imagining Identity: Nos/otras (Us/Other) and the New Tribalism." Draft manuscript, 2003.

———. "Haciendo caras, una entrada." Anzaldúa, *Making Face, Making Soul/Haciendo Caras* xv–xxviii.

———. *Interviews/Entrevistas.* Ed. AnaLouise Keating. New York: Routledge, 2000.

———. "La Prieta." Moraga and Anzaldúa, *This Bridge Called My Back* 198–209.

———. "Let Us Be the Healing of the Wound: the Coyolxauhqui imperative—la sombra y el sueno." *One Wound for Another/ Una herida por otra: Testimonios de latin@s in the U.S. through Cyberspace (11 septiembre 2001–11 marzo 2002).* Ed. Clara Lomas and Claire Joysmith. Mexico City: Centro de Investigaciones Sobre América del Norte (CISAN), at the Universidad Nacional Autónoma de México (UNAM), 2003.

———, ed. *Making Face, Making Soul/Haciendo Caras: Creative and Critical Perspectives by Women of Color.* San Francisco: Aunt Lute Foundation, 1990.

———. "now let us shift . . . the path of conocimiento . . . inner work, public acts." Anzaldúa and Keating, *this bridge we call home* 540–78.

———. "To(o) Queer the Writer—*Loca, escritora y chicana*." *Inversions: Writing by Dykes, Queers, and Lesbians*. Ed. Betsy Warland. Vancouver: P Gang, 1991. 249–64.

———. "(Un)natural bridges, (Un)safe spaces." Anzaldúa and Keating, *this bridge we call home* 1–5.

Anzaldúa, Gloria E. and AnaLouise Keating, eds. *this bridge we call home: radical visions for transformation*. New York: Routledge, 2002.

Aparicio, Frances. "On Multiculturalism and Privilege: A Latina Perspective." *American Quarterly* 46.4 (1994): 575–88.

Appiah, Kwame Anthony. "The Conservation of 'Race.'" *Black American Literature Forum* 23 (1989): 37–60.

———. "The Uncompleted Argument: Du Bois and the Illusion of Race." Gates, *"Race," Writing, and Difference* 21–37.

Apple, Michael W. "Foreword." Kincheloe et al., *White Reign* ix–xiii.

———. "Freire and the Politics of Race in Education." *International Journal of Leadership in Education* 6.2 (2003): 107–18.

Awiakta, Marilou. *Selu: Seeking the Corn-Mother's Wisdom*. Golden, CO: Fulcrum, 1993.

Babb, Valerie. *Whiteness Visible: The Meaning of Whiteness in American Literature and Culture*. New York: New York UP, 1998.

Bailey, Alison. "Locating Traitorous Identities: Toward a View of Privilege—Cognizant White Character." *Hypatia* 13 (1998): 27–43.

Baker, Houston, Jr. "Caliban's Triple Play." Gates, *"Race," Writing, and Difference* 381–95.

Bakhtin, Mikhail. "Discourse in the Novel." *The Dialogic Imagination*. Ed. Michael Holquist. Trans. Caryl Emerson and Michael Holquist. Austin: U of Texas P, 1981. 259–422.

———. *Speech Genres & Other Late Essays*. Trans. Vern W. McGee. Ed. Caryl Emerson and Michael Holquist. Austin: U of Texas P, 1986.

———. *Toward a Philosophy of the Act*. Ed. Vadim Liapunov and Michael Holquist. Trans. Vadim Liapunov. Austin: U of Texas P, 1993.

Baldwin, James. "On Being 'White' . . . and Other Lies." Roediger, *Black on White* 177–80.

———. *The Price of the Ticket: Collected Nonfiction 1948–1985*. New York: St. Martin's, 1985.

Barber, Benjamin R. "The Educated Student: Global Citizen or Global Consumer?" *Liberal Education* (Spring 2002): 22–28.

Bay, Mia. *The White Image in the Black Mind: African-American Ideas about White People, 1830–1925*. New York: Oxford UP, 2000.

Bell, Derrick. *Faces at the Bottom of the Well: The Permanence of Racism*. New York: Basic Books, 1992.

Bennett, Lerone, Jr. *Before the Mayflower: A History of Black America*. 1966. New York: Penguin, 1988.

Bennett, Paula. "The Pea that Duty Locks: Lesbians and Feminist-Heterosexual Readings of Emily Dickinson's Poetry." *Lesbian Texts and Contexts: Radical Revisions*. Ed. Karla Jay and Joanne Glasgow. New York: New York UP, 1990. 104–25.

Bercovitch, Sacvan. "Afterword." *Ideology and Classic American Literature*. Ed. Sacvan Bercovitch and Myra Jehlen. Cambridge: Cambridge UP, 1986. 418–42.

Birtha, Becky. *For Nights Like This One*. San Francisco: Frog in the Well P, 1983.

Blumenfeld, Warren J., ed. *Homophobia: How We All Pay the Price*. Boston: Beacon, 1992.

Bonnett, Alastair. "Anti-racism and the Critique of 'White' Identities." *New Community* 22 (1996): 97–110.

———. "Constructions of 'Race,' Place and Discipline: Geographies of 'Racial' Identity and Racism." *Ethnic and Racial Studies* 19 (1996): 864–83.

———. "Constructions of Whiteness in European and American Anti-Racism." Torres et al., *Race, Identity, and Citizenship* 200–18.

Bonnett, Alastair. "Geography, 'Race' and Whiteness: Invisible Traditions and Current Challenges." *Area* 29.3 (1997): 193–99.

Bornstein, Kate. *Gender Outlaw: On Men, Women, and the Rest of US*. New York: Vintage, 1995.

Brannon, Lil and Brenda M. Greene, ed. *Rethinking American Literature*. Urbana, IL: NCATE, 1997.

Bridges, Flora Wilson. *Resurrection Song: African-American Spirituality*. Maryknoll, NY: Orbis Books, 2001.

Brodkin, Karen. *How Jews Became White Folks and What that Says about Race in America*. New Brunswick, NJ: Rutgers UP, 1999.

Brown, Rita Mae. *Rubyfruit Jungle*. 1973. New York: Bantam, 1983.

Brubaker, Roger, Mara Loveman, and Peter Stamatov. "Ethnicity as Cognition." *Theory and Society* 33 (2004): 31–64.

Bruce-Novoa, Juan. "Canonical and Noncanonical Texts: A Chicano Case Study." *Redefining American Literary History*. Ed. A. LaVonne Ruoff and Jerry Washington Ward. New York: Modern Languages Association, 1990.

Butler, Octavia. *Adulthood Rites*. New York: Warner, 1988.

———. *Clay's Ark*. 1984. New York: Warner, 1996.

———. *Dawn*. New York: Warner, 1987.

———. *Fledgling*. New York: Seven Stories Press, 2005.

———. *Imago*. New York: Warner, 1989.

———. *Kindred*. 1979. Boston: Beacon, 1988.

———. *Mind of My Mind*. 1977. New York: Avon, 1978.

———. *Parable of the Sower*. New York: Four Walls Eight Windows, 1993.

———. *Parable of the Talents*. New York: Four Walls, 1998.

———. *Patternmaster*. 1976. New York: Warner, 1995.

———. *Survivor*. Garden City, NY: Doubleday, 1978.

———. *Wild Seed*. 1980. New York: Warner, 1988.

Cajete, Gregory. *Look to the Mountain: An Ecology of Indigenous Education*. Skyland, NC: Kivaki P, 1994.

———. *Native Science: Natural Laws of Interdepencence*. Santa Fe: Clear Light Publishers, 2000.

Canaan, Andrea. "Brownness." Moraga and Anzaldúa, *This Bridge Called My Back* 232–37.

Caputi, Jane. "Interview With Paula Gunn Allen." *Trivia* 16 (1990): 50–67.

Chambers, Ross. "The Unexamined." Hill, *Whiteness* 187–203.

Castillo, Ana. *Loverboys*. New York: Penguin, 1996.

———. *The Mixquiahuala Letters*. Houston: Bilingual P, 1986.

———. *Sapagonia*. Houston: Bilingual P, 1990.

———. *So Far from God*. 1993. New York: Penguin, 1994.

Cather, Willa. *Sapphira and the Slave Girl*. 1940. New York: Vintage, 1975.

Cervenak, Sarah J., et. al. "Imagining Differently: The Politics of Listening in a Feminist Classroom." Anzaldúa and Keating, *this bridge we call home* 341–56.

Chân Không. *Learning True Love: How I Learned and Practiced Social Change in Vietnam*. Berkeley, CA: Parallax P, 1993.

Chávez, Denise. *The Last of the Menu Girls*. Houston: Arte Público, 1986.

Chickering, Arthur W. "Curricular Content and Powerful Pedagogy." *Encouraging Authenticity and Spirituality in Higher Education*. Ed. Arthur W. Chickering, Jon C. Dalton, and Liesa Stamm. San Francisco, CA: Jossey-Bass, 2006. 113–44.

Christian, Barbara. "The Race for Theory." Anzaldúa, *Making Face, Making Soul/Haciendo Caras* 335–45.

Collins, Patricia Hill. *Black Feminist Thought: Knowledge, Consciousness, and the Politics of Empowerment*. Rev. Ed. New York: Routledge, 2000.

———. *From Black Power to Hip Hop: Racism, Nationalism, and Feminism*. Philadelphia: Temple UP, 2006.

Cone, James H. "Afterword." *A Black Theology of Liberation*. 1970. Twentieth Anniversary Edition. New York: Orbis, 1996.197–201.

Courvant, Diana. "Speaking of Privilege." Anzaldúa and Keating, *this bridge we call home* 458–63.

Crenshaw, Kimberlé Williams. "Color-Blind Dreams and Racial Nightmares: Reconfiguring Racism in the Post–Civil Rights Era." *Birth of a Nation 'hood: Gaze, Script, and Spectacle in the O.J. Simpson Case*. Ed. Toni Morrison and Claudia Brodsky Lacour. New York: Pantheon, 1997. 97–168.

cunningham, e. christi. "The 'Racing' Cause of Action and the Identity Formally Known as Race: The Road to Tamazunchale." *Rutgers Law Review* 30 (1999): 709–30.

Davidson, Arnold I. "Sex and the Emergence of Sexuality." *Critical Inquiry* 14 (1987): 16–48.

de la Peña, Terri. *Margins*. Seattle: Seal P, 1992.

DeLillo, Don. *White Noise*. New York: Penguin, 1985.

Deloria, Vine, Jr. *God Is Red: A Native View of Religion*. Rev. Ed. Goldon, CO: Fulcrum, 1994.

Dillard, Cynthia B. *On Spiritual Strivings: Transforming an African American Woman's Academic Life*. Albany: SUNY P, 2006.

Doane, Ashley W. and Eduardo Bonilla-Silva, eds. *White Out: The Continuing Significance of Racism*. New York: Routledge, 2003.

Doane, Woody. "Rethinking Whiteness Studies." Doane and Bonilla-Silva, *White Out* 3–18.

Delgado, Richard. *The Rodrigo Chronicles: Conversations about America and Race*. New York: New York U, 1995.

———. "Storytelling for Oppositionists and Others: A Plea for Narrative." *Michigan Law Review* 87 (1989): 2411, 2413–14.

de Lauretis, Teresa. "Queer Theory: Lesbian and Gay Sexualities, An Introduction." *differences* 3 (1991): iii–xviii.

Dollimore, Jonathan. *Sexual Dissidence: Augustine to Wilde, Freud to Foucault*. New York: Oxford UP, 1991.

Downer, Christa. "The Making of a Pluralistic Egalitarian Society: Reconceptualizing the Rhetoric of Multiculturalism for the 21st Century." Diss., Texas Woman's U, 2006.

Dwyer, Owen J. and John Paul Jones III. "White Socio-Spatial Epistemology." *Social & Cultural Geography* 1.2 (2000): 209–22.

Dykewomon, Elana. "The Body Politic—Mediations on Identity." Anzaldúa and Keating, *this bridge we call home* 599–612.

Dyson, Michael Eric. "The Labor of Whiteness, the Whiteness of Labor." Torres et al., *Race, Identity, and Citizenship* 219–24.

Dyer, Richard. "White." *Screen: The Journal for Education in Film and Television* 29 (1988): 44–64.

Ebony. The White Problem in America. Chicago: Johnson Publishing Company, 1966.

Ellison, Ralph. *Invisible Man*. 1947. New York: Vintage, 1990.

Emerson, Ralph Waldo. 1836. *Nature*. In *Emerson: Essays & Poems*. New York: Library of America, 1996. 7–49.

Faderman, Lillian. *Surpassing the Love of Men: Romantic Friendship and Love between Women from the Renaissance to the Present*. New York: William Morrow, 1981.

Fauset, Jessie Redmon. 1929. *Plum Bun: A Novel without a Moral*. Boston: Beacon P, 1999.

Fernandes, Leela. *Transforming Feminist Practice: Non-Violence, Social Justice and the Possibilities of a Spiritualized Feminism*. San Francisco: Aunt Lute Books, 2003.

Fernández, Carlos A. "La Raza and the Melting Pot: A Comparative Look at Multiethnicity." *Racially Mixed People in America*. Ed. Maria P. P. Root. Newbury Park, CA: Sage, 1992. 126–43.

Fetterley, Judith and Marjorie Pryse, eds. *American Women Regionalists, 1850–1910*. New York: W.W. Norton, 1995.

Findlen, Barabra, ed. *Listen Up! Voices from the Next Feminist Generation*. 2nd ed. New York: Seal P, 2001.

Fishkin, Shelley Fisher. "Interrogating 'Whiteness,' Complicating 'Blackness': Remapping American Culture." *Criticism and the Color Line: Desegregating American Literary Studies*. Ed. Henry Wonham. New Brunswick: Rutgers UP, 1996. 251–90.

Ford, Richard T. "Race As Culture? Why Not?" *UCLA Law Review* 47 (2000): 1803–15.

Foster, Frances Smith. "But Is It Good Enough to Teach?" Brannon and Greene, *Rethinking American Literature* 193–202.

Fox, Matthew. *Original Blessing*. Santa Fe: Bear & Co., 1988.

Francis, Becky. "Commonalities and Difference? Attempts to Escape from Theoretical Dualisms in Emancipatory Research in Education." *International Studies in Sociology of Education* 11.2 (2001): 157–72.

Frankenberg, Ruth, ed. *Displacing Whiteness: Essays in Social and Cultural Criticism*. Durham: Duke UP, 1997.

———. *The Social Construction of Whiteness: White Women, Race Matters*. Minneapolis: U of Minnesota P, 1993.

Freire, Paulo. *Pedagogy of the Oppressed*. 1970. Trans. Myra Bergman Ramos. New York: Continuum, 1990.

Friedman, Thomas L. *The World Is Flat: A Brief History of the Twenty-First Century*. New York: Farrar, Strauss, & Giroux, 2005.

Frye, Marilyn. "On Being White: Toward a Feminist Understanding of Race and Race Supremacy." *The Politics of Reality: Essays in Feminist Theory*. Freedom, CA: Crossing P, 1983. 110–27.

———. "White Woman Feminist." *Willful Virgin: Essays in Feminism 1976–1992*. Freedom, CA: Crossing P, 1992. 147–69.

———. *Willful Virgin: Essays in Feminism, 1976–1992*. Freedom, CA: Crossing P, 1992.

Fuss, Diana. *Inside/Out: Lesbian Theories, Gay Theories*. New York: Routledge, 1991.

Gallagher, Charles A. "Redefining Racial Privilege in the United States." *The New Jersey Project Journal* 8 (1997): 28–39.

———. "Researching Race, Reproducing Racism." *Review of Education, Pedagogy & Cultural Studies* 21.2 (1999): 165–91.

———. "White Reconstruction in the University." *Socialist Review* 94 (1995): 165–87.

Garber, Linda, ed. *Tilting the Tower: Lesbian/Teaching in the Queer Nineties*. New York: Routledge, 1994.

Gates, Henry Louis, Jr. *Loose Canons: Notes on the Culture Wars*. New York: Oxford UP, 1992.

———, ed. *"Race," Writing, and Difference*. Chicago: U of Chicago P, 1985.

———. "Writing 'Race' and the Difference It Makes." Gates, *"Race," Writing, and Difference* 1–20.

Ginzberg, Ruth. "Audre Lorde's (Nonessentialist) Lesbian Eros." *Hypatia* 7 (1992): 73–90.

Giroux, Henry. "Democracy and the Discourse of Cultural Difference: Towards a Politics of Border Pedagogy." *British Journal of Sociology of Education* 12 (1991): 501–20.

———. "Post-Colonial Ruptures and Democratic Possibilities: Multiculturalism as Anti-Racist Pedagogy." *Cultural Critique* 21 (1992): 5–39.

———. "Racial Politics, Pedagogy, and the Crisis of Representation in Academic Multiculturalism." *Social Identities* 6.4 (December 2000): 493–510.

———. "Rewriting the Discourse of Racial Identity: Towards a Pedagogy and Politics of Whiteness." *Harvard Educational Review* 67.2 (1997): 275–320.

Goldberg, David. *Racist Culture: Philosophy and the Politics of Meaning*. Cambridge: Blackwell, 1993.

Grahn, Judy. *The Queen of Swords*. Boston: Beacon P, 1987.

———. *The Work of a Common Woman*. Trumansburg, NY: Crossing P, 1978.

Grasso, Linda. *The Artistry of Anger: Black and White Women's Literature in America, 1820 to 1860*. U of North Carolina P, 2002.

Greenberg, David F. *The Construction of Homosexuality*. Chicago: U of Chicago P, 1988.

Green, Jamison. *Becoming a Visible Man*. Nashville, TN: Vanderbilt UP, 2004.

Grobman, Laurie. "Toward a Multicultural Pedagogy: Literary and Nonliterary Traditions." *MELUS* 26 (2001): 221–40.

Guillaumin, Collette. "Race and Nature: The System of Marks." Trans. Mary Jo Lakeland. *Feminist Issues* 8 (1988): 25–43.

Gwaltney, John Langston. *Drylongso: A Self-Portrait of Black America*. 1980. New York: New P, 1993.

Hacking, Ian. "Why Race Still Matters." *Daedalus* 134.1 (Winter 2005): 102–16.

Hale, Grace Elizabeth. *Making Whiteness: The Culture of Segregation in the South, 1890–1940*. New York: Pantheon Books, 1998.

Hall, Donald E. *Queer Theories*. New York: Palgrave Macmillan, 2003.

Hall, Kim F. "Beauty and the 'Beast' of Whiteness: Teaching Race and Gender." *Shakespeare Quarterly* 47:4 (1996): 461–75.

Hall, Stuart. "Cultural Identities and Diaspora." *Identity: Community, Culture, Difference*. Ed. Jonathan Rutherford. London: Lawrence & Wishart, 1990. 222–37.

———. "Racist Ideologies and the Media." *Media Studies*. 2nd ed. Ed. Paul Marris and Sue Thornham. Washington Square, NY: New York UP. 271–82.

Haney-López, Ian. "Community Ties and Law School Faculty Hiring: The Case for Professors Who Don't Think White." *Beyond a Dream Deferred: Multicultural Education and the Politics of Excellence*. Ed. Becky W. Thompson and Sangeeta Tyagi. Minneapolis: U of Minnesota P, 1993. 100–30.

———. "Race on the 2010 Census: Hispanics and the Shrinking White Majority." *Daedalus* 134.1 (2005): 42–52.

———. *White By Law: The Legal Construction of Race*. New York: New York UP, 1996.

Haraway, Donna. *Primate Visions: Gender, Race, and Nature in the World of Modern Science*. New York: Routledge, 1989.

Harjo, Joy. *The Good Luck Cat*. Ill. Paul Lee. New York: Harcourt Children's Books, 2000.

———. "Introduction." *Reinventing the Enemy's Language: Contemporary Native Women's Writings of North America*. Ed. Joy Harjo and Gloria Bird. New York: W.W. Norton, 1997. 19–31.

———. *The Spiral of Memory: Interviews*. Ed. Laura Coltelli. Ann Arbor: U of Michigan P, 1996.

Harris, E. Lynn. *And This Too Shall Pass*. New York: Doubleday, 1996.

———. *Invisible Life*. New York: Doubleday, 1991.

———. *Just As I Am*. New York: Doubleday, 1994.

Harrison-Kahan, Lori. " 'Queer myself for good and all': *The House of Mirth* and the Fictions of Lily's Whiteness." *Legacy* 21.1 (2004) 34–49.

Hartigan, John, Jr. "Establishing the Fact of Whiteness." *American Anthropologist* 99.3 (1997): 495–505.

———. *Racial Situations: Class Predicaments of Whiteness in Detroit*. Princeton: Princeton UP, 1999.

Hattam, Victoria. "Ethnicity and the Boundaries of Race: Rereading Directive 15." *Daedalus* 134.1 (2005): 61–69.

Hayward, Jeremy. *Perceiving Ordinary Magic: Science and Intuitive Wisdom*. Boston: Shambala, 1984.

Hemphill, Essex, ed. *Brother to Brother: New Writings by Black Gay Men*. Boston: Alyson, 1991.

Hernández-Ávila, Inés. "In the Presence of Spirit(s): A Meditation on the Politics of Solidarity and Transformation." Anzaldúa and Keating, *this bridge we call home* 530–38.

Hesse, Barnor. "It's Your World: Discrepant M/Multiculturalisms." *Social Identities* 3.3 (1997): 375–94.

Hill, Mike. *After Whiteness: Unmaking an American Majority*. New York: New York UP, 2004.

———, ed. *Whiteness: A Critical Reader*. New York: New York UP, 1997.

Hinojosa, Rolando. *Klail City*. Houston: Arte Público P, 1987.

His Holiness the Dalai Lama. *Ethics for the New Millenium*. New York: Riverhead Books, 1999.

Hollinger, David A. *Postethnic America: Beyond Multiculturalism*. New York: Basic Books, 1995.

hooks, bell. *Black Looks: Race and Representation*. Boston: South End P, 1992.

———. "Buddhism and the Politics of Domination." *Mindful Politics: A Buddhist Guide to Making the World a Better Place*. Ed. Melvin McLeod. Boston: Wisdom Publications, 2006. 57–62.

———. *Feminist Theory: From Margin to Center*. Boston: South End P, 1984.

———. *Talking Back: Thinking Feminist, Thinking Black*. Boston: South End P, 1989.

———. *Yearning: Race, Gender, and Cultural Politics*. Boston: South End P, 1990.

Hopkins, Pauline. *Contending Forces: A Romance Illustrative of Negro Life North and South*. 1900. New York: Oxford UP, 1991.

Hughes, Langston. *The Ways of White Folks*. 1933. New York: Vintage, 1971.

Hurtado, Aída. "The Trickster's Play: Whiteness in the Subordination and Liberation Process." Torres et al., *Race, Identity, and Citizenship* 225–43.

Ignatiev, Noel. *How the Irish Became White*. New York: Routledge, 1996.

Jacobson, Matthew Frye. *Whiteness of a Different Color: European Immigrants and the Alchemy of Race*. Cambridge: Harvard UP, 1998.

Jaimes, M. Annette, ed. *The State of Native America: Genocide, Colonization, and Resistance*. Boston: South End P, 1992.

James, Joy. "The Academic Addict: Mainlining (and Kicking) White Supremacy (WS)." Yancy, *What White Looks Like* 263–67.

JanMohamed, Abdul R. "The Economy of Manichean Allegory: The Function of Racial Difference in Colonialist Literature." Gates, *"Race," Writing, and Difference* 78–106.

Jarrett, Gene Andrew, ed. *African American Literature beyond Race; A Reader*. New York: New York UP, 2006.

Jarrett, Gene Andrew. "Judging a Book by Its Writer's Color." *The Chronicle of Higher Education* 52 (July 28, 2006): B12.

———. "Not Necessarily Race Matter." Jarrett, *African American Literature beyond Race* 1–22.

Jay, Gregory. "The End of 'American' Literature: Toward a Multicultural Practice." *College English* 53 (1991): 264–81.

Jen, Gish. *Typical American*. New York: Penguin Books, 1991.

Jewett, Sarah Orne. *The Country of the Pointed Firs and Other Stories*. 1986. New York: Signet, 2000.

Jordan, Winthrop. *The White Man's Burden: Historical Origins of Racism in the United States*. Oxford: Oxford UP, 1974.

Kaplan, E. Ann. "The 'Look' Returned: Knowledge Production and Constructions of 'Whiteness' in Humanities Scholarship and Independent Film." Hill, *Whiteness* 316–28.

Kay, Jackie. *Trumpet*. New York: Vintage, 2000.

Keating, AnaLouise. "Shifting Perspectives: Spiritual Activism, Social Transformation, and the Politics of Spirit." *EntreMundos/AmongWorlds: New Perspectives on Gloria E. Anzaldúa*. Ed. AnaLouise Keating. New York: Palgrave Macmillan, 2005. 241–54.

———. "Transcendentalism Then and Now: Towards a Dialogic Theory and Praxis of Multicultural US Literature." Brannon and Greene, *Rethinking American Literature* 50–68.

———. *Women Reading Women Writing: Self-Invention in Paula Gunn Allen, Gloria Anzaldúa, and Audre Lorde*. Philadelphia: Temple UP, 1996.

———. "Writing, Politics, and *las Lesberadas*: Platicando con Gloria Anzaldúa." *Frontiers: A Journal of Women Studies* 14 (1993): 105–30.

Kenan, Randall. *Let the Dead Bury the Dead*. New York: Harcourt Brace, 1992.

———. *Visitation of Spirits*. New York: Grove P, 1989.

Kincheloe, Joe L. "The Struggle to Define and Reinvent Whiteness: A Pedagogical Analysis." *College Literature* 26 (Fall 1999): 162–94.

Kincheloe, Joe L. and Shirley R. Steinberg. "Addressing the Crisis of Whiteness." Kincheloe et al., *White Reign* 3–29.

Kincheloe, Joe L., Shirley R. Steinberg, Nelson M. Rodriguez, and Ronaled E. Chennault. *White Reign: Deploying Whiteness in America*. New York: St. Martin's P, 1998.

King, Robert H. *Thomas Merton and Thich Nhat Hanh: Engaged Spirituality in an Age of Globalization*. New York: Continuum, 2001.

King, Thomas. *The Truth about Stories: A Native Narrative*. 2003. Minneapolis: U of Minnesota P, 2005.

Kroeber, Karl. "An Interview with Jack Salzman, Director of the Columbia University Center for American Culture Studies." *American Indian Persistence and Resurgence*. Ed. Karl Kroeber. Durham: Duke UP, 1994. 50–57.

Lakoff, Robin Tolmach. *The Language War*. Berkeley: U of California P, 2000.

Larsen, Nella. *Quicksand and Passing*. 1928 and 1929. Ed. Deborah McDowell. New Brunswick, NJ: Rutgers UP, 1986.

Laszlo, Ervin. *The Connectivity Hypothesis: Foundations of an Integral Science of Quantum, Cosmos, Life, and Consciousness*. Buffalo: State U of New York P, 2003.

Lee, Chang-Rae. *Native Speaker*. New York: Riverhead Books, 1995.

Leonardo, Zeus. "The Souls of White Folk: Critical Pedagogy, Whiteness Studies, and Globalization Discourse." *Race, Ethnicity, and Education* 5.1 (2002): 29–50.

Lewis, Amanda E. "Some Are More Equal than Others: Lessons on Whiteness from School." Doane and Bonilla-Silva, *White Out* 159–72.

Lionnet, Françoise. *Autobiographical Voices: Race, Gender, Self-Portraiture*. Ithaca: Cornell UP, 1989.

Lipsitz, George. *The Possessive Investment in Whiteness: How White People Profit from Identity Politics*. Philadelphia: Temple UP, 1998.

———. "The White 2K Problem." *Cultural Values* 4.2 (2000): 518–24.

Lorde, Audre. *Sister Outsider: Essays and Speeches*. Freedom, CA: Crossing P, 1984.

———. *Zami: A New Spelling of My Name*. Freedom, CA: The Crossing P, 1982.

Loy, David R. *The Great Awakening: A Buddhist Social Theory*. Boston: Wisdom Publications, 2003.

Lugones, María. "Playfulness, 'World'-Travelling, and Loving Perception." Anzaldúa, *Making Face, Making Soul/Haciendo Caras* 390–402.

Lugones, María and Elizabeth V. Spelman. "Have We Got a Theory for You! Feminist Theory, Cultural Imperialism, and the Demand for 'the Woman's Voice.'" *Hypatia Reborn: Essays in Feminist Philosophy*. Ed. Azizah Y. Al-Hibri and Margaret A. Simons. Bloomington, IN: Indiana UP, 1990. 18–33.

Mahoney, Martha R. "Segregation, Whiteness, and Transformation." 1995. *Critical White Studies: Looking Behind the Mirror*. Ed. Richard Delgado and Jean Stefancic. Philadelphia: Temple UP, 1997. 654–57.

———. "The Social Construction of Whiteness." 1995 *Critical White Studies: Looking Behind the Mirror*. Ed. Richard Delgado and Jean Stefancic. Philadelphia: Temple UP, 1997. 330–33.

Maitino, John R. and David R. Peck. "Introduction." *Teaching American Ethnic Literatures: Nineteen Essays*. Ed. John R. Maitino and David R. Peck. Albuquerque: U of New Mexico P, 1996. 3–16.

Mani, Lata. *Interleaves: Ruminations on Illness and Spiritual Life*. Northampton, MA: Interlink Publishing Group, 2001.

Marcus, Jane. *Hearts of Darkness: White Women Write Race*. New Brunswick, NJ: Rutgers UP, 2004.

Marshall, Paule. *Praisesong for the Widow*. New York: Plume, 1992.

Mayberry, Katherine J, ed. *Teaching What You're Not: Identity Politics in Higher Education*. New York: New York UP, 1996.

McLaren, Peter. "Multiculturalism and the Postmodern Critique: Toward a Pedagogy of Resistance and Transformation." *Between Borders: Pedagogy and the Politics of Cultural Studies*. Ed. Henry Giroux and Peter Mclaren. New York: Routledge, 1994. 192–222.

————. *Revolutionary Multiculturalism: Pedagogies of Dissent for the New Millennium.* Boulder, CO: Westview P, 1997.

McLaren, Peter and Tomaz Tadeu da Silva. "Decentering Pedagogy: Critical Literacy, Resistance and the Politics of Memory." *Paulo Freire: A Critical Encounter.* Ed. Peter McLaren and Peter Leonard. New York: Routledge, 1993. 47–83.

McLendon, John H., III. "On the Nature of Whiteness and the Ontology of Race: Toward a Dialectical Materialist Analysis." Yancy, *What White Looks Likes* 211–25.

McPhail, Mark Lawrence. "Complicity: The Theory of Negative Difference." *Howard Journal of Communications* 3.1 and 2 (1991): 1–13.

————. "The Politics of Complicity: Second Thoughts about the Social Construction of Racial Equality." *Quarterly Journal of Speech* 80 (1994): 343–81.

Mercer, Kobena. "Skin Head Sex Thing: Racial Difference and the Homoerotic Imaginary." *How Do I Look? Queer Film and Video.* Ed. Bad Object-Choices. Seattle: Bay P, 1991. 169–210.

Miles, Robert. "Apropos the Idea of 'Race' . . . Again." *Theories of Racism: A Reader.* Ed. Les Back and John Solomos. London and New York: Routledge, 2000. 123–43.

Miles, Robert and Rudolfo D. Torres. "Does 'Race' Matter?" Torres et al., *Race, Identity, and Citizenship* 19–38.

Miller, Wayne Charles. "Cultural Consciousness in a Multicultural Society: The Uses of Literature." *MELUS* 8 (1981): 24–44.

Mills, Charles W. "White Supremacy as Sociopolitical System: A Philosophical Perspective." Doane and Bonilla-Silva, *White Out* 35–48.

Mitchell, Bonnie and Joe R. Feagin. "America's Racial-Ethnic Cultures: Opposition within a Mythical Melting Pot." *Toward the Multicultural University.* Ed. Benjamin P. Bowser, Terry Jones, and Gale Auletta Young. Westport, CT: Praeger, 1995. 65–86.

Mohanty, Chandra. *Feminism without Borders: Decolonizing Theory, Practicing Solidarity.* Durham, NC: Duke UP, 2002.

————. "On Race and Voice: Challenges for Liberal Education in the 1990s." *Cultural Critique* 14 (Winter 1989–90): 179–208.

Momaday, N. Scott. *House Made of Dawn.* New York: Harper & Row, 1966.

Montoya, Margaret E. "Celebrating Racialized Legal Narratives." *Crossroads, Directions, and a New Critical Race Theory.* Ed. Francisco Valdes, Jerome McCristal Culp, and Angela P. Harris. Philadelphia: Temple UP, 2002. 243–50.

Moon, Michael. *Disseminating Whitman: Revision and Corporeality in Leaves of Grass.* Cambridge: Harvard UP, 1991.

Moraga, Cherríe. "Algo secretamente amado." *Third Woman: The Sexuality of Latinas* 4 (1989): 151–56.

————. *Giving Up the Ghost.* Los Angeles: West End P, 1986.

————. "La Güera." Moraga and Anzaldúa, *This Bridge Called My Back* 27–34.

————. *Loving in the War Years: lo que nunca pasó por sus labios.* Boston: South End P, 1983.

Moraga, Cherríe and Gloria Anzaldúa, eds. 1981. *This Bridge Called My Back: Writings by Radical Women of Color.* New York: Kitchen Table: Women of Color P, 1983.

Morgan, Kelly. "Our Place in World Literature." *Reclaiming the Vision Past, Present, and Future: Native Voices for the Eighth Generation.* Ed. Lee Francis and James Bruchac. Greenfield Center, NY: Greenfield Review P, 1996. 30–31.

Mori, Toshio. *Yokohama, California.* 1949. Seattle: U of Washington P, 1993.

Morrison, Toni. Interview with Elizabeth Farnsworth. *News Hour with Jim Lehrer.* PBS. March 1998.

————. *Paradise.* New York: Alfred A. Knopf, 1998.

————. *Playing in the Dark: Whiteness and the Literary Imagination.* Cambridge: Harvard UP, 1992.

———. *Sula*. New York: Penguin, 1973.

Mosha, R. Sambuli. "The Inseparable Link between Intellectual and Spiritual Formation in Indigenous Knowledge and Education: A Case Study in Tanzania?" *What Is Indigenous Knowledge? Voices from the Academy*. Ed. Ladislaus M. Semali and Joe L. Kincheloe. New York/London: Farmer, 1999. 209–25.

Muhanji, Cherry. *Her*. San Francisco: Aunt Lute, 1990.

Myers, Linda J. et al. "Identity Development and Worldview: Toward an Optimal Conceptualization." *Journal of Counseling & Development* 70 (1991): 54–63.

Myers, Linda James. *Understanding an Afrocentric World View: Introduction to an Optimal Psychology*. 1988. Dubuque, IA: Kendall/Hunt Publishing, 1993.

Nguyen, Alexander. "The Souls of White Folk." *The American Prospect* 11.17 (2000): 46–49.

Nhat Hanh, Thich. *Interbeing: Fourteen Guidelines for Engaged Buddhism*. 1987. 3rd ed. Berkeley: Parallax P, 1998.

———. *Transformation at the Base: Fifty Verses on the Nature of Consciousness*. Berkeley, CA: Parallax P, 2001.

Nielsen, Aldon Lynn. *Reading Race: White American Poets and the Racial Discourse in the Twentieth Century*. Athens: U of Georgia P, 1988.

Norton, Jody. "Transchildren, Changelings, and Fairies: Living the Dream and Surviving the nightmare in Contemporary America." Anzaldúa and Keating, *this bridge we call home* 145–54.

Norwood, Kimberly Jade. "The Virulence of Blackthink and How Its Threat of Ostracism Shackles Those Deemed Not Black Enough." *Kentucky Law Journal* 93 (2004/2005): 146–99.

Oliver, Kelly. "Identity, Difference, and Abjection." *Theorizing Multiculturalism: A Guide to the Current Debate*. Ed. Cynthia Willett. Oxford: Blackwell, 1998. 169–86.

———. *Witnessing: Beyond Recognition*. Minneapolis: U of Minnesota P, 2001.

Omi, Michael and Howard Winant. *Racial Formation in the United States from the 1960s to the 1980s*. 2nd ed. New York: Routledge, 1994.

Ortiz, Fernando. *Cuban Counterpoint: Tobacco and Sugar*. Trans. Harriet de Onís. New York: Alfred A. Knopf, 1947.

Ortiz, Simon. *The People Shall Continue*. Ill. Sharol Graves. Rev. ed. San Francisco: Children's Book P, 1988.

Owens, Louis. *Mixedblood Messages: Literature, Film, Family, Place*. Norman: U of Oklahoma P, 1998.

Palumbo-Liu, David. "Introduction." *The Ethnic Canon: Histories, Institutions, and Interventions*. Ed. David Palumbo-Liu. Minneapolis: U of Minnesota P, 1995. 1–27.

Paredes, Patricia Palacios. "Note: Latinos and the Census: Responding to the Race Question." *The George Washington Law Review* 75 (2005): 146–63.

Parker, Robert Dale. "Material Choices: American Fictions, the Classroom, and the Post-Canon." *American Literary History* 5 (1993): 89–110.

Patterson, Orlando. *The Ordeal of Integration: Progress and Resentment in America's "Racial" Crisis*. Washington, DC: Civitas/Counterpoint, 1997.

Perry, Pamela. "White Means Never Having to Say You're Ethnic." *Journal of Contemporary Ethnography* 30.1 (2001): 56–91.

Pfeil, Fred. *White Guys: Studies in Postmodern Dominance and Difference*. New York: Routledge, 1995.

Pharr, Suzanne. *Homophobia: A Weapon of Sexism*. Little Rock, AR: Chardon P, 1988.

Piercy, Marge. *Woman on the Edge of Time*. 1976. New York: Ballantine Books, 1997.

Pineda, Cecile. *Face*. New York: Viking, 1985.

Piper, Adrian. "Passing for White, Passing for Black." *New Feminist Criticism: Art, Identity, Action*. Ed. Joanna Frueh, Cassandra L. Langer, and Arlene Raven. New York: HarperCollins, 1994. 216–47.

Pollock, Mica. "Race Wrestling: Struggling Strategically with Race in Educational Practice and Research." *American Journal of Education* 111 (2004): 25–67.

Ponce, Mary Helen. *The Wedding*. Houston: Arté Público, 1989.

powell, john a. "The Colorblind Multiracial Dilemma: Racial Categories Reconsidered." *University of San Francisco Law Review* 31 (1997): 789–807.

———. "Disrupting Individualism and Distributive Remedies with Intersubjectivity and Empowerment: An Approach to Justice and Discourse." *University of Maryland Law Journal of Race, Religion, Gender and Class* 1 (2001): 1–23.

———. "Does Living a Spiritually Engaged Life Mandate Us to Be Actively Engaged in Issues of Social Justice?" *University of St. Thomas Law Journal: Fides et Iustitia* 1 (2003): 30–38.

———. "A Minority–Majority Nation: Racing the Population in the Twenty-First Century." *Fordham Urban Law Journal* 29 (2002): 1395–416.

powell, john a. "The Multiple Self: Exploring between and beyond Modernity and Postmodernity." *Minnesota Law Review* 81 (June 1997): 1481–521.

———. "Our Private Obsession, Our Public Sin: The 'Racing' of American Society: Race Functioning as a Verb before Signifying as a Noun." *Law and Inequality Journal* 15 (1997): 99–126.

———. "Whiteness: Some Critical Perspectives: Dreaming of a Self beyond Whiteness and Isolation." *Washington University Journal of Law & Policy* 18 (2005): 14–145.

———. "Whites Will Be Whites: The Failure to Interrogate Racial Privilege." *University of San Francisco of Law* 34 (Spring 2000): 419–65.

Powell, Timothy B. "All Colors Flow into Rainbows and Nooses: The Struggle to Define Academic Multiculturalism." *Cultural Critique* 55 (2003): 152–81.

———. *Ruthless Democracy: A Multicultural Interpretation of the American Renaissance*. Princeton: Princeton UP, 2000.

Prashad, Vijay. *Everybody Was Kung Fu Fighting: Afro-Asian Connections & the Myth of Cultural Purity*. Boston: Beacon, 2001.

Pratt, Minnie Bruce. "Identity: Skin Blood Heart." *Yours in Struggle*. Ed. Elly Bulkin, Minnie Bruce Pratt, and Barbara Smith. 1984. Ithaca, NY: Firebrand Books, 1988.

Prewitt, Kenneth. "Racial Classification in America: Where Do We Go from Here?" *Daedalus* 134.1 (2005): 5–17.

Pruett, Christina. "The Complexions of 'Race' and the Rise of 'Whiteness' Studies." *Clio* 32.1 (2002): 25–52.

Queen, Carol and Laurence Schimel, eds. *Pomosexuals: Challenging Assumptions about Sexuality and Gender*. San Francisco: Cleis P, 1997.

Ratcliffe, Krista. "Eavesdropping as Rhetorical Tactic: History, Whiteness, and Rhetoric." *JAC: Journal of Advanced Composition* 20 (2000): 101–34.

Rich, Adrienne. "Compulsory Heterosexuality and Lesbian Existence." *Signs* 5 (1980): 631–60.

Rich, Marc D. and Aaron Castelan Cargile. "Beyond the Breach: Transforming White Identities in the Classroom." *Race, Ethnicity, and Education* 7.4 (2004): 251–65.

Robinson, Reginald Leamon. " 'Expert' Knowledge: Introductory Comments on Race Consciousness." *Third World Law Journal* 20 (1999): 145–83.

———. "Human Agency, Negated Subjectivity, and White Structural Oppression: An Analysis of Critical Race Practice/Praxis." *American University Law Review* 53 (August 2004): 1361–420.

———. "Poverty, the Underclass, and the Role of Race Consciousness: A New Age Critique of Black Wealth/White Wealth and American Apartheid." *Indiana Law Review* 34 (2001): 1378–443.

———. "Race Consciousness: A Mere Means of Preventing Escapes from the Control of Her White Masters?" *Touro Law Review* 15.2 (1999): 401–38.

———. "Race, Myth and Narrative in the Social Construction of the Black Self." *Howard Law Journal* 40 (1996): 1–148.

———. "Split Personalities: Teaching and Scholarship in Nonstereotypical Areas of the Law." *Western New England Law Review* 19 (1997): 73–79.

Roediger, David, ed. *Black on White: Black Writers on What It Means to Be White*. New York: Schocken, 1998.

——. "Critical Studies of Whiteness, USA: Origins and Arguments." *Theoria: A Journal of Social and Political Theory* 98 (2001): 72–98.

——. "Is There a Healthy White Personality?" *Counseling Psychologist* 27 (1999): 239–44.

——. *Toward the Abolition of Whiteness*. New York: Verso, 1994.

Rosenberg, Marshall B. "Spiritual Basis of Nonviolent Communication: A Question and Answer Session with Marshall B. Rosenberg, Ph.D." <http://www.cnvc.org/spirital.htm>. March 22, 2006.

Rothenberg, Paula. "The Construction, Deconstruction, and Reconstruction of Difference." *Hypatia* 5 (1990): 42–57.

Rubin, Henry. *Self-Made Men: Identity and Embodiment among Transsexual Men*. Nashville, TN: Vanderbilt UP, 2003.

Russ, Joanna. *The Female Man*. 1975. Boston: Beacon P, 2000.

Sampson, Robert J. and Stephen W. Raudenbush. "Seeing Disorder: Neighborhood Stigma and the Social Construction of 'Broken Windows.' " *Social Psychology Quarterly* 67.4 (2004): 319–43.

Sandoval, Chela. "Theorizing White Consciousness for a Post-Empire World: Barthes, Fanon, and the Rhetoric of Love." Frankenberg, *Displacing Whiteness: Essays in Social and Cultural Criticism* 87–106.

——. "U.S. Third World Feminism: The Theory and Method of Oppositional Consciousness in the Postmodern World." *Genders* 10 (Spring 1991): 1–25.

Sartwell, Crispin. *Act Like You Know: African-American Autobiography and White Identity*. Chicago: U of Chicago P, 1998.

Scheurich, James Joseph. "Toward a White Discourse on White Racism." *Educational Researcher* 22 (1993): 5–10.

Scheurich, James J. and Michelle D. Young. "Coloring Epistemologies: Are Our Research Epistemologies Racially Biased?" *Educational Researcher* 26 (1997): 4–16.

Schuyler, George. *Black No More*. 1931. New York: Random House, 1999.

Sedgwick, Eve Kosofsky. *Between Men: English Literature and Male Homosocial Desire*. New York: Columbia UP, 1985.

——. *Epistemology of the Closet*. Berkeley: U of California P, 1990.

——. *Tendencies*. Chapel Hill: U of North Carolina P, 1993.

Semali, Ladislaus M. and Joe L. Kincheloe. "Introduction: What Is Indigenous Knowledge and Why Should We Study It?" *What Is Indigenous Knowledge? Voices from the Academy*. Ed. Ladislaus M. Semali and Joe L. Kincheloe. New York/London: Farmer, 1999. 3–57.

Shanley, Kathryn W. "The Indians America Loves to Love and Read: American Indian Identity and Cultural Appropriation." *American Indian Quarterly* 21 (1997): 675–702.

Shipler, David K. *Country of Strangers: Blacks and Whites in America*. New York: Alfred A. Knopf, 1997.

Shohat, Ella and Robert Stam. *Unthinking Eurocentrism: Multiculturalism and the Media*. New York: Routledge, 1994.

Silko, Leslie Marmon. *Ceremony*. New York: Penguin, 1977.

Sinclair, April. *Ain't Gonna Be the Same Fool Twice*. New York: Hyperion, 1996.

——. *Coffee Will Make You Black*. New York: Hyperion, 1994.

Smith, Barbara. "Towards a Black Feminist Criticism." 1977. *The New Feminist Criticism: Essays on Women, Literature, Theory*. New York: Pantheon, 1985.

——. "The Truth that Never Hurts: Black Lesbians in Fiction in the 1980s." *Feminisms*. Ed. Robyn R. Warhol and Diane Price Herndl. New Brunswick: Rutgurs UP, 1991. 690–712.

Smith, David Lionel. "What Is Black Culture?" *The House that Race Built: Black Americans, U.S. Terrain*. Ed. Wahneema Lubiano. New York: Pantheon, 1997. 178–94.

Snavely, Cynthia A. "Native American Spirituality: Its Use and Abuse by Anglo-Americans." *Journal of Religious & Theological Information* 14 (2001): 91–103.

Srivastava, Sarita. " 'You're calling me a racist?' The Moral and Emotional Regulation of Antiracism and Feminism." *Signs* 31.1 (2005): 29–62.

Steinberg, Stephen. *The Ethnic Myth: Race, Ethnicity, and Class in America.* 1982. Boston: Beacon, 1989.

Stockton, Sharon. " 'Blacks vs. Browns': Questioning the White Ground." *College English* 57 (1995): 166–81.

Stoltenberg, John. *Refusing to Be a Man: Essays on Sex and Justice.* New York: Penguin, 1989.

Sue, Derald Wing. "Whiteness and Ethnocentric Monoculturalism: Making the 'Invisible' Visible." *American Psychologist* 59 (November 2004): 761–69.

Sundquist, Eric J. *To Wake the Nations: Race in the Making of American Literature.* Cambridge: Harvard UP, 1993.

Tatum, Beverly Daniel. "Changing Lives, Changing Communities: Building a Capacity for Connection in a Pluralistic Context." *Education as Transformation: Religious Pluralism, Spirituality, and a New Vision for Higher Education in America.* Ed. Victor H. Kazanjian, Jr. and Peter L. Laurence. New York: Peter Lang, 2000. 79–88.

Taylor, Paul C. "Silence and Sympathy: Dewey's Whiteness." Yancy, *What White Looks Like* 227–41.

Thandeka. *Learning to Be White: Money, Race, and God in America.* New York: Continuum, 2000.

Thompson, Audrey. "Entertaining Doubts: Enjoyment and Ambiguity in White, Antiracist Classrooms." *Passion and Pedagogy: Relation, Creation, and Transformation in Teaching.* Ed. Elijah Mirochnik and Debora C. Sherman. New York: Peter Lang, 2002. 431–52.

Thornton, Michael C. "Is Multiracial Status Unique? The Personal and Social Experience." *Racially Mixed People in America.* Ed. Maria P. P. Root. Newbury Park, CA: Sage, 1992. 321–25.

Tompkins, Jane. *Sensational Designs: Cultural Work of American Fiction 1790–1860.* Oxford UP, 1985.

TuSmith, Bonnie. *All My Relatives: Community in Contemporary Ethnic American Literatures.* Ann Arbor: U of Michigan P, 1994.

Trinh T. Minh-ha. *When the Moon Waxes Red: Representation, Gender and Cultural Politics.* New York: Routledge, 1991.

Torres, Rudolfo D., Louis F. Mirón, and Jonathan Xavier Inda. "Introduction." Torres et al., *Race, Identity, and Citizenship* 1–16.

———, eds. *Race, Identity, and Citizenship.* Malden, MA: Blackwell Publishers, 1999.

Valerio, Max Wolf. "Now that You're a White Man': Changing Sex in a Postmodern World—Being, Becoming, and Borders." Anzaldúa and Keating, *this bridge we call home* 239–54.

Violet, Indigo. "Linkages: A Personal–Political Journey with Feminist of Color Politics." Anzaldúa and Keating, *this bridge we call home* 651–63.

Wagner, Sally Roesch. *Sisters in Spirit: Haudenosaunee (Iroquois) Influence on Early American Feminists.* Summertown, TN: Book Publishers Company, 2001.

Walker, Alice. *The Color Purple.* New York: Washington Square, 1982.

———. *Now Is the Time to Open Your Heart.* New York: Random, 2004.

———. *Possessing the Secret of Joy.* New York: Harcourt Brace Jovanovich, 1992.

———. *The Temple of My Familiar.* New York: Harcourt Brace Jovanovich, 1989.

Wallace, David L. "Out in the Academy: Heterosexism, Invisibility, and Double Consciousness." *College English* 65.1 (2002):53–66.

Weaver, Jace. *That the People Might Live: Native American Literatures and Native American Community.* New York: Oxford UP, 1997.

Webster, Yehudi O. *The Racialization of America.* New York: St. Martin's P, 1992.

West, Cornel. *Prophetic Reflections: Notes on Race and Power in America.* Monroe, ME: Common Courage, 1993.

West, Dorothy. *The Living Is Easy*. 1948. New York: Feminist P, 1982.

Wiegman, Robyn. "Whiteness Studies and the Paradox of Particularity." *Boundary 2* 26.3 (1999): 115–50.

Wilchins, Riki Anne. *Read My Lips: Sexual Subversion and the End of Gender*. Milford, CT: Firebrand Books, 1997.

Williams, Kim M. "Multiracialism and the Civil Rights Future." *Daedalus* 134 (2005): 53–57.

Williams, Patricia. *The Alchemy of Race and Rights: Diary of a Law Professor*. Cambridge: Harvard UP, 1991.

———. *Seeing a Color-Blind Future: The Paradox of Race*. New York: Farrar, Straus & Giroux, 1997.

Williams, Walter L. *The Spirit and the Flesh: Sexual Diversity in American Indian Culture*. 1986. Boston: Beacon, 1992.

Willie, Sarah Susannah. *Acting Black: College, Identity, and the Performance of Race*. New York: Routledge, 2002.

Wittig, Monique. *The Straight Mind and Other Essays*. Boston: Beacon, 1992.

Wray, Matt and Annalee Newitz, eds. *White Trash: Race and Class in America*. New York: Routledge, 1997.

Yamashita, Karen Tei. *Through the Arc of the Rain Forest*. Minneapolis, MN: Coffee House P, 1990.

Yancy, George. "Introduction: Fragments of a Social Ontology of Whiteness." Yancy, *What White Looks Like* 1–23.

Yancy, George, ed. *What White Looks Like: African-American Philosophers on the Whiteness Question*. New York: Routledge, 2004.

York, Michael. "New Age Commodification and Appropriation of Spirituality." *Journal of Contemporary Religion* 16 (2001): 361–72.

Yoshihara, Mari. *Embracing the East: White Women and American Orientalism*. New York: Oxford UP, 2002.

Zack, Naomi, ed. *American Mixed Race: The Culture of Microdiversity*. Lantham, Maryland: Rowman & Littlefield, 1995.

———. "Preface." Zack, *American Mixed Race* ix–xi.

———. *Race and Mixed Race*. Philadelphia: Temple UP, 1993.

———. *Thinking About Race*. Belmont, CA: Wadsworth, 1998.

Zackodnik, Teresa. "Fixing the Color Line: The Mulatto, Southern Courts, and Racial Identity." *American Studies Quarterly* 53 (2001): 420–51.

ACKNOWLEDGMENTS

Writing this book has been a collective endeavor, and I have many people to thank. I'm grateful to my parents, Tom and Joann Keating, for instilling in me such a great love of reading and for providing me with an excellent education. I'm grateful to my teachers and professors, from pre-K through graduate school, especially those like Gary Masquelier who refused to take the easy answers and insisted that I push my thinking further. I am grateful to my family—Eddy Lynton and Jamitrice Keating-Lynton—for the many daily lessons our lives together create. You have been generous, supportive, and wise; I am very blessed to have you in my life.

I am grateful to all of my students, past and present. Your eagerness to learn, your challenging questions, and your provocative comments teach me so much! Thanks to Glenda Lehrmann, F.I.R.S.T. librarian at Texas Woman's University, for tracking down many sources for me. Thanks to Claire L. Sahlin, ideal colleague and program director extraordinaire.

I am grateful to Jacqui Alexander, Gloria Evangelina Anzaldúa, Sue Beyerlein, Mary Loving Blanchard, Renae Bredin, Christa Downer, Angela Johnson Fisher, Carrie McMaster, Harry McMaster, Nery Morales, Timothy Powell, and Jean Wyatt for reading parts of *Teaching Transformation*. Your thoughtful comments have greatly strengthened this book. A special thanks to Jo-Ann Stankus for reading the entire manuscript and for sending me useful articles and quotations. Thanks to Nadine Barrett, Patricia Stukes, and Doreen Watson for insightful conversations and much-needed laughter about our classroom work. Mil gracias to Gloria Anzaldúa for our many conversations on teaching and 'race' and for encouraging me to take risks. Comadre, you are sorely missed!

I am grateful to the authors who taught me how to teach: Gloria Anzaldúa, Henry Giroux, bell hooks, Maria Lugones, Peter McLaren, and Paula Rothenberg. Special thanks to Octavia Butler, whose novels taught me how to teach 'race.' Thanks also to the writers who inspire, nurture, and challenge me: Paula Gunn Allen, M. Jacqui Alexander, James Baldwin, Ralph Waldo Emerson, Leela Fernandes, Joy Harjo, Patricia Hill Collins, June Jordan, Audre Lorde, Jane Roberts, Chela Sandoval.

I am grateful to Palgrave Macmillan, especially Amanda Moon, Emily Leithauser, and Kristy Lilas, and to Newgen Imaging Systems, especially Maran Elancheran, for seeing this manuscript into print. Portions of several chapters were previously published in earlier versions, and I gratefully acknowledge permission to republish. Part of chapter two was published in a different form as " 'Making New Connections': Transformational Multiculturalism in the Classroom" in *Pedagogy* 4.1 (2004): 93–117; an earlier version of chapter three was published as "Interrogating 'Whiteness,' (De)Constructing 'Race' " in *College English* 57 (1995): 901–18; an earlier version of chapter four appeared as "Exposing 'Whiteness,' Unreading 'Race': (De)Racialized Reading Tactics in the Classroom" in *Reading Sites: Social Differences and Reader Response*, coedited by Elizabeth Flynn and Patsy Schweickart (New York: MLA, 2004), 314–42; and an earlier version of part of chapter five was published as "Heterosexual Teacher, Lesbian/Gay/ Bisexual Text: Teaching the Sexual *Other*" in *Tilting the Tower: Lesbian/ Teaching in the Queer Nineties*, edited by Linda Garber (New York: Routledge, 1994), 97–107. Although they might not recognize parts of my arguments, given the ways my ideas have developed and changed over the years, I am grateful to these editors for their comments on early drafts of this project.

And, finally, I am grateful to the orishas, espiritus, and ancestors for guiding me, whispering words of encouragement that nourish my body/heart/mind/spirit and inspire my vision.

INDEX